Communications in Computer and Information Science   670

Commenced Publication in 2007
Founding and Former Series Editors:
Alfredo Cuzzocrea, Dominik Ślęzak, and Xiaokang Yang

More information about this series at http://www.springer.com/series/7899

Robin Doss · Selwyn Piramuthu
Wei Zhou (Eds.)

Future Network Systems and Security

Second International Conference, FNSS 2016
Paris, France, November 23–25, 2016
Proceedings

 Springer

Editors
Robin Doss
School of Information Technology
Deakin University
Burwood, VIC
Australia

Selwyn Piramuthu
Department of Information Systems
  and Operations Management
University of Florida, Warrington College
  of Business Administration
Gainesville, FL
USA

Wei Zhou
Information and Operations Management
  Department
ESCP Europe
Paris
France

ISSN 1865-0929   ISSN 1865-0937 (electronic)
Communications in Computer and Information Science
ISBN 978-3-319-48020-6   ISBN 978-3-319-48021-3 (eBook)
DOI 10.1007/978-3-319-48021-3

Library of Congress Control Number: 2016954932

Printed on acid-free paper

This Springer imprint is published by Springer Nature
The registered company is Springer International Publishing AG
The registered company address is: Gewerbestrasse 11, 6330 Cham, Switzerland

Preface

Welcome to the proceedings of the Future Network Systems and Security Conference 2016 held in Paris, France!

The network of the future is envisioned as an effective, intelligent, adaptive, active, and high-performance Internet that can enable applications ranging from smart cities to tsunami monitoring. The network of the future will be a network of billions or trillions of entities (devices, machines, things, vehicles) communicating seamlessly with one another, and it is rapidly gaining global attention from academia, industry, and government. The main aim of the FNSS conference series is to provide a forum that brings together researchers from academia, practitioners from industry, standardization bodies, and government to meet and exchange ideas on recent research and future directions for the evolution of the future Internet. The technical discussions were focused on the technology, communications, systems, and security aspects of relevance to the network of the future.

We received paper submissions by researchers from around the world including the USA, Australia, New Zealand, Germany, India, Taiwan, Japan, Sweden, UK, and UAE among others. After a rigourous review process 12 full papers and one short paper were accepted covering a wide range of topics. The overall acceptance rate for the conference was 35 % ensuring that the accepted papers were of a very high quality. We thank the Technical Program Committee for their hard work in ensuring such an outcome.

November 2016

Robin Doss
Selwyn Piramuthu
Wei Zhou

Organization

FNSS 2016 was held at ESCP Europe, Paris Campus, during November 23–25, 2016.

Conference Chairs

Robin Doss   Deakin University, Australia
Selwyn Piramuthu   University of Florida, USA
Wei Zhou   ESCP Europe, France

Technical Program Committee

Maythem Abbas	Universiti Teknologi Petronas, Malaysia
S. Agrawal	Delhi Technological University (DTU), India
Rana Khudhair Ahmed	Al-Rafidain University College, Iraq
Adil Al-Yasiri	University of Salford, UK
Abdul Halim Ali	Universiti Kuala Lumpur – International College, Malaysia
Elizabeth Basha	University of the Pacific, USA
Aniruddha Bhattacharjya	Guru Nanak Institute of Technology (GNIT), India
David Boyle	Imperial College London, UK
Doina Bucur	University of Groningen, The Netherlands
Bin Cao	Harbin Institute of Technology, P.R. China
Yue Cao	University of Surrey, UK
Arcangelo Castiglione	University of Salerno, Italy
Sammy Chan	City University of Hong Kong, Hong Kong, SAR China
Kesavaraja D.	Dr. Sivanthi Aditanar College of Engineering, India
Eleonora D'Andrea	University of Pisa, Italy
Soumya Kanti Datta	EURECOM, France
Safiullah Faizullah	Hewlett-Packard, USA
Stephan Flake	Redknee Germany OS GmbH, Germany
Miguel Franklin de Castro	Federal University of Ceará, Brazil
Felipe Garcia-Sanchez	Universidad Politecnica de Cartagena (UPCT), Spain
Razvan Andrei Gheorghiu	Politehnica University of Bucharest, Romania
Mikael Gidlund	Mid Sweden University, Sweden
Andrzej Glowacz	AGH University of Science and Technology, Poland
Shweta Jain	York College CUNY, USA
Manoj Joy	Amal Jyothi College of Engineering, India
Hussain Mohammed Dipu Kabir	The Hong Kong University of Science and Technology, Hong Kong, SAR China
Mounir Kellil	CEA LIST, France
Piotr Korbel	Lodz University of Technology, Poland

Contents

AuthentIx: Detecting Anonymized Attacks via Automated Authenticity Profiling

Mordechai Guri[✉], Matan Monitz[✉], and Yuval Elovici[✉]

Cyber-Security Research Center, Ben-Gurion University of the Negev, Beer-Sheva, Israel
{gurim,monitz,elovici}@post.bgu.ac.il

Abstract. In the modern era of cyber-security attackers are persistent in their attempts to hide and mask the origin of their attacks. In many cases, attacks are launched from spoofed or unknown Internet addresses, which makes investigation a challenging task. While *protection* from anonymized attacks is an important goal, *detection* of anonymized traffic is also important in its own right, because it allows defenders to take necessary preventative and defensive steps at an early stage, even before the attack itself has begun. In this paper we present AuthentIx, a system which measures the authenticity of the sources of Internet traffic. In order to measure the authenticity of traffic sources, our system uses passive and active profiling techniques, which are employed in both the network and the application protocols. We also show that performing certain cross-views between different communications layers can uncover inconsistencies and find clients which are suspicious. We present our system design and describe its implementation, and evaluate AuthentIx on traffic from authentic and non-authentic sources. Results show that our system can successfully detect anonymous and impersonated attackers, and furthermore, can be used as a general framework to cope with new anonymization and hiding techniques.

Keywords: Anonymization · Attacks · IP profiling · Proxy · VPN · Onion routing

1 Introduction

In today's cyber-security arena, adversaries continuously try to breach a wide range of industrial, financial, and governmental organizations for sabotage and espionage purposes. The motivation for preventing any cyber-attack is obvious, but tracing an attack's origin (mainly its Layer 3 IP address) is a crucial element of the investigation of cyber incidents. Detecting an attack's origin may be required in order to effectively implement network based counterattacks, take legal action, and conduct kinetic attacks. Such actions may be taken by capable defenders equipped with direct and specific knowledge of the attacker's identity and location [1]. Attack attribution also allows the defender to gain insight into attackers' intentions and methodologies by enabling the attribution of multiple attacks - seemingly originating from different sources - to a single attacker. Attackers, on the other hand, aim to thwart attribution efforts in order to protect themselves from being identified and continue conducting attacks unimpeded. Anonymization is one of the major attack methodologies used to hide the origin of an attack, and

© Springer International Publishing AG 2016
R. Doss et al. (Eds.): FNSS 2016, CCIS 670, pp. 1–11, 2016.
DOI: 10.1007/978-3-319-48021-3_1

attackers use various services and network technologies to accomplish this and mask their identity [2, 3]. Previous research discusses attack attribution and methods of defeating an attackers' anonymity [1]. However most existing methods focus on a specific protocol layer (e.g., TCP/IP, HTTP/S, Tor) or require resources and methodologies that are not readily available to most defenders (e.g., access to traffic or logs from multiple nodes along the attackers' path).

1.1 Client Profiling

In this paper we present a system for authenticating the source of network traffic by estimating whether specific network traffic originates from an authentic source. The system will measure and estimate the authenticity, and will assign a low authenticity grade to traffic originating from clients that hide their identity behind anonymization proxies or VPN services. In addition, clients that try to hide their identities, disguise their operating system, browser type, or network timing are considered more suspicious. We present the design, architecture, implementation, and evaluation of AuthentIx, a framework which inspects the traffic between Internet services and clients and profiles the clients based on their authenticity level. Our system explores different communication layers and performs differential and cross-layer analysis to detect suspicious clients. To the best of our knowledge, our system is the first to present an architecture and implementation of multi-layers and extendible system for evaluation of network traffic authenticity.

The rest of this paper is structured as follows: Sect. 2 surveys anonymization methods. Section 3 presents our system's design and architecture. Section 4 discusses the evaluation of the system and presents our results. Related work is included in Sect. 5. We provide a conclusion in Sect. 6.

2 Anonymization Methods

In this section we survey the main methods used in order to gain anonymity on the Internet. It's important to note that anonymity by itself doesn't necessarily reflect a client's malicious intent. Anonymity has been used for many purposes including privacy protection, to bypass restrictions and censorship, and so on. However, anonymization when coupled with other behavior may lead to the suspicion of particular clients. For example, malformed TCP/IP packets repeatedly sent from an anonymized source may be an initial sign of a server attack or DDoS campaign.

2.1 Tor Onion Routing

Onion routing [4], developed at the U.S. Naval Research Laboratory, is a technique for anonymous communication over the Internet. In onion routing, traffic is encrypted and encapsulated in layers (like the layers of an onion). The traffic is routed through different network nodes referred to as 'onion routers.' At each node, the next destination is determined by 'peeling' a single layer of the traffic. At the final hope, the message arrives at its destination and is decrypted. During routing, the sender remains anonymous, because each node is only aware of the previous node and the one that follows it in the routing path.

2.2 HTTP(S) Proxies

HTTP(S) proxies [5] enable the anonymization of web-surfing traffic by establishing the HTTP(S) connections from different IP addresses which masks the originator. The destination server (the server that handles the web request) only sees requests from the proxy server, which hides information about the end user's IP address. There are many free and commercial HTTP proxy services for anonymization or traffic acceleration [6, 7].

2.3 SOCKS Proxies

A SOCKS proxy [8] is a server that establishes a TCP or UDP connection to another server on behalf of a client. The SOCKS server does not interfere with the traffic content. Web browsers and other types of clients (e.g., torrent) can be configured to talk to a server via a SOCKS server. As HTTP proxies, SOCKS proxies can mask the original IP address for anonymization purposes.

2.4 WEB/CGI Proxies

WEB/CGI proxies [9, 10] are used when it's impossible to use standard HTTP(S)/ SOCKS proxies. These servers run a proxy in the form of a CGI script. They retrieve HTTP(S) content and change all of the links on the page so that they go through this server script and subsequently return the changed page to the user's side. In this way, the user is kept as anonymous as possible from any servers.

2.5 VPN Protocols (PPTP, SSL-VPN, IPSEC-VPN)

A Virtual Private Network (VPN) is a technology that allows a secure network connection over a non-secure or public network such as the Internet. A typical VPN situation is when an organization wants to link two sites together. Users should be able to connect to all endpoints (computers, printers, etc.) on both sites as if they were all on a single local network. There are several VPN tunneling protocols, handling and encapsulating the traffic at different layers. IP security (IPSec) is used to secure communication over the Internet. Secure Sockets Layer (SSL) and Transport Layer Security (TLS) use symmetric cryptography to establish secure communications over the Internet. Point-To-Point Tunneling Protocol (PPTP) is another tunneling protocol used to implement a private network over the Internet by encapsulating data link (PPP) packets. In the same way, Layer 2 Tunneling Protocol (L2TP) is a protocol used to tunnel layer 2 communication traffic between two sites over the Internet within UDP packets. Working with VPNs, users are able to hide their original address from the destination: a user first connects to a remote VPN server and then manages its connections from there. Since VPNs work at the TCP/IP layer, they fit any application protocol such as HTTP, Torrent, FTP, and SMTP. Today there are many VPN services available which offer privacy and anonymization by allowing users to route their traffic through different locations (e.g., other countries) [11–13].

2.6   General Tunneling Interfaces (SSH, GRE, and IPv6 Tunnels)

Tunneling protocols are used to encapsulate different protocols, tunneling the traffic to a remote server. For example, SSH tunnels consist of an encrypted tunnel created through an SSH protocol connection. Such a tunnel can be used to transfer unencrypted traffic over a network through an encrypted channel. The client's computer connects to a remote SSH server which in turn manages the connection with the destination. This way the source IP address and ports of the client are kept hidden from the server. Such tunneling protocols have been used for anonymization purposes [14], and other tunneling protocols include the GRE and IPv6 tunnels.

Table 1 summarizes various anonymization and hiding methods, the layers they are operatings, the anonymity level and the transperancy to the user. As can be seen, the onion routing has the highest anonymity level since even the onion routers are not aware of the source and destination address. The application proxies provide a medium level of anonymization. In these methods, the network level identifiers (IP, ports) are masked from the target server. However, some application identifiers (e.g., browser cookies) may be leaked. As can be seen, VPN and tunneling interfaces only mask identifiers at the network levels, while all application level identifiers are transferred as is. At the transperancy and usability, the Tor and most of the proxy techniques are not transparent to the end user (browsing can be interrupted), and the onion routing usually requires the use of a special application. In addition, when an HTTPS proxy is being used, the user may be alerted about the use of unknown certificates. Finally, VPN and tunneling interfaces are most transparent to the user, since they work at the low layers of the network protocols.

Table 1.   Various anonymization and hiding methods

Method	Layers	Anonymity	Transparency
Tor Onion Routing	Application	High	Low
HTTP(S) proxy	Application	Medium	Low
WEB/CGI proxy	Application	Medium	Low
SOCKS PROXY	Application	Medium	Medium
VPN	Network	Network Level	High
Tunneling Interfaces	Network	Network Level	High

3   System Design and Architecture

We designed AuthentIx, an extendable system which inspects network traffic and assigns a grade, which represents the 'level' of authenticity of the clients (the connection peers).

The AuthentIx internal engine receives the content of the traffic, analyzes the network and application layers and provides cross-layer validation. There are two types of traffic data that can be given as input to the AuthentIx engine: (1) passive sniffing of the network traffic from a defender's perspective, and (2) network traffic with active participation in communication (e.g., port scanning). The AuthentIx engine can query third party sources unrelated to the traffic source. The external database includes geographic IP databases, blacklists, DNS information, and more. Figure 1 presents the architecture of the AuthentIx engine.

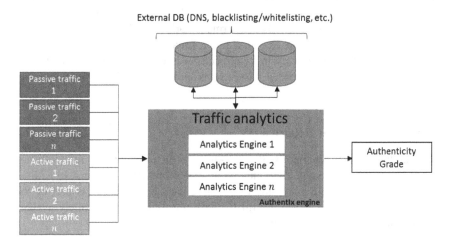

Fig. 1. The AuthentIx system's modular architecture.

The AuthentIx engine receives different traffic recordings in PCAP format as input. The data includes the traffic information from the IP layer to the application layer (e.g., HTTP). Various analytics engines handle and process the traffic, and give it an authenticity grade. The output of the AuthentIx engine is the list of authenticity grades. Our implementation includes seven analytics engines. In the following subsection we briefly describe these engines and their algorithms.

3.1 TCP ACK RTT/HTTP GET Request Time Interval Ratio

This analytic engine attempts to identify timing differences between application-level delay and network-level delay for each TCP connection. The round-trip time (RTT) for a TCP connection is calculated using its initial SYN and ACK packets. This parameter is compared to the delay between consecutive HTTP requests which is generated by the client's browser or attack tools. The higher this ratio is, the more likely that there is a discrepancy – that the endpoint which the TCP session is held with is not the endpoint that authentically originated the HTTP request. For example, if a user hides behind an application level proxy (e.g., HTTP proxy), it's likely to find a discrepancy between the HTTP and TCP timing. Such a deviation indicates an instance of application level anonymity.

3.2   Cross-Checking User-Agent and Passive Fingerprint Based OSs

Each HTTP request contains a user-agent string which provides information about the requesting endpoint, including its operating system (OS) and version. The OS discovered by the user-agent string is compared to a guess made by p0f [15] - a tool that passively analyses the traffic and determines the operating system behind the network-level session. A mismatch between the two indicates a non-authentic session. For example, some attack tools which spoof the OS field in the user-agent, will be detected by this method. This method will also detect some anonymity HTTP(S) proxies which override the OS information in the request headers. A cross-check between the passive and active fingerprints of the OS is also performed. Various methods (usually JavaScript based techniques) are used to provide an active estimation of the client OS. E.g., there are several online tools which try to fingerprint an Internet user by various browser methods [16]. A mismatch between passive and active fingerprinting may indicate a non-authentic session.

3.3   Inconsistency in the Reported OS

If the operating system reported by the user-agent is inconsistent, this may indicate a proxy server that serves many endpoints with several operating systems. In addition, some attack tools generate automatic or multiple user-agent strings as part of their attacks. This method is primarily useful for detecting traffic from publicly available anonymization proxies and VPNs providers.

3.4   TCP MSS and Encapsulated Message Size

The maximum segment size (MSS) is a TCP session parameter that provides details about the maximum packet size allowed for the TCP session and is derived from the type of physical connection, layers 2–3 header size, and encapsulation overhead. 1460 is a typical MSS value which is commonly seen in the Internet. When all or some of the traffic is encapsulated, a TCP stack in the routing path may decrease the MSS size for the next route. This is done because the full packet is bigger than the encapsulated one. A lower than normal MSS value in a TCP session may indicate an additional encapsulation overhead resulting from a non-authentic session.

3.5   External Database Querying

In this analytic engine the source of the traffic is checked against external databases. During our tests we used MaxMind [17], a proxy identification service which is used for online fraud detection applications. This service provides a proxy score to help identify IP addresses that are considered high risk. Various IP reputation and DNS blacklist databases [18, 19] are also queried by this analytic engine. These services and databased are continuously updated with IP and domain names which are known to be the source of DDoS or other types of attack campaigns. The Tor exist node list [20] is also queried to see if an address is part of the Tor network. The Tor project reports all exit nodes' addresses, for a variety of purpose (Tor is not designed to hide the fact that a user is connected to the Tor network, but to protect his original address). IP geolocation databases [21] were used in order to identify the origin of source addresses. This information can be cross-checked with data from the HTTP headers (e.g., language). Along with the scores listed above, extended session information is extracted using WHOIS queries [22]. This information sometimes includes the DNS owner's address and contact information.

3.6   Port Generation Analysis

The operating system allocates TCP/UDP port numbers when a session is intitiated. Some operating systems generate a list of the ports sequentially. Based on the assumption that the TCP/UDP stack of proxy servers is more active than that of a legitimate user's computer, sequential connections from the same user will not expose sequential numbers. This analytic engine analyzes traffic and the outgoing and incoming TCP and UDP ports used over time and between sessions. If port numbers do not exhibit standard behavior, it is an indication that proxy or anonymization servers have been used.

3.7   Network Distance

This engine simply measures the average number of hops calculated by the received packet's Time to Leave (TTL) field. This information can be used to estimate the distance from a real client. In addition, using active methods of tracerouting and the response time can be used to determine whether the network distance is authentic.

Table 2 summarizes the analytic engines used in our AuthentIx implementation and their relevance to various anonymization methods.

Table 2. Analytic engine used in our implementation

Method	Detection Provided
Request Time Interval Ratio	Proxies, anonymity services
Cross-Checking User-Agent OS	Fake clients, anonymity services, proxies, attack tools
Inconsistency in Reported OS	Fake clients, anonymity services, proxies, attack tools
TCP MSS and Encapsulated Message Size	VPNs, tunneling services
External Database Querying	Proxies, anonymity services, attacking IPs, VPNs, Tor
Port Generation Analysis	–
Network Distance	–

4   Evaluation

We implemented and evaluated AuthentIx with the seven analytic engines presented in the previous section. For the evaluation, we created anonymized clients by using various anonymization and hiding techniques presented in Sect. 2 and used them to browse a simple web page on an evaluation server while we recorded the network traffic. Our anonymized clients used the following methods: (1) Tor browsing, (2) HTTP proxy, (3) WEB/CGI proxy, (4) SOCK proxy, (5) VPN, and (6) L2 tunneling services. We used both public and private VPN services. We analyzed the traffic flow with AuthentIx. Tor anonymized clients were successfully detected by AuthentIx using the published exit address list. As for clients hiding behind HTTP proxies, the timing methods and OS detection based methods proved very effective in the detection of proxy usage. In some cases, HTTP proxies can also be detected by the existence of specific headers that may even reveal the original source address. Clients communicating through WEB/CGI proxies were successfully detected by the same methods that were effective with HTTP proxies. VPN tunnels were successfully detected using the various OS fingerprinting methods and to a greater extent using the MSS analysis method. When using a VPN, the client sends encrypted packets to the VPN server, using approximately 150 bytes for an encapsulated header. This creates decapsulated packets of 1350 bytes which are sent to the defended server. Therefore, TCP sessions with an MSS value of 1400 or less are suspected of originating from a VPN. The tunneling methods are subject to the same overhead issue as the VPN servers and were successfully detected using the same method. External sources such as proxy score and blacklists showed limited results, detecting only the publicly available (free or fee-based) proxies and VPN services. These services will unlikely be able to detect a server purchased by an attacker for the purpose of anonymization. A summary of the major results is provided in Table 3.

Table 3. Summary of the evaluation results

Method	Timing analysis	OS fingerprint analysis	Encapsulation overhead analysis	3rd party blacklists and databases	TCP source port analysis
Tor	Partial	Yes	–	Yes	No
HTTP proxy	Yes	Yes	No	Partial	No
WEB/CGI proxy	Yes	Yes	No	Partial	No
SOCKS PROXY	Yes	Yes	No	Partial	No
Tunneling and VPN	No	Yes	Yes	Partial	No

AuthentIx can be used to generate an authenticity score based on the combine results from various analytic engines. Figure 2 illustrates web-based output generated by AuthentIx based on a combination of eight analytic engines. The results are presented in a visual web interactive interface which consists of a map with all source IPs and DNSs appears in the traffic. A click on specific IP address opens a dialog with detailed evaluation results regard this specific source address.

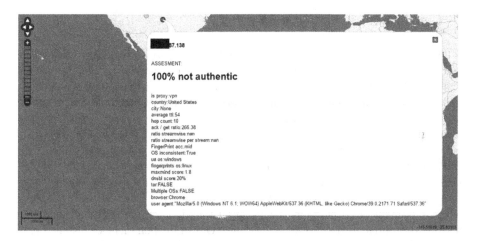

Fig. 2. AutenthIx Web Interface: Proxy detected by the system based on timing parameters.

5   Related Work

Over the years a variety of concepts for anonymization on the Internet have been introduced. Koukis et al. [23] presented a generic anonymization framework for network traffic using a so-called anonymization application programming interface. The concept of onion routing for traffic anonymization was first discussed by Goldschlag [24] and Reed [25] and analyzed by Syverson [26]. Using anonymization proxies for privacy was discussed in [27–29]. VPN Tunnels, proxies, and onion routing for identity hiding were overviewed in [30]. An analytical review of methods of providing Internet anonymity was provided by Savchenko and Gatsenko [31] and Farah [32]. While there have been a wide range of studies on anonymization of Internet traffic, detecting and tracing anonymized connections has received less attention from security researchers. Coull et al. discussed different techniques for evaluating anonymized network data [33]. Various attacks on Tor anonymization for adversaries who can observe traffic at different points of the communication path were presented in [34, 35]. Liming discussed passive and active fingerprinting methods to identify TLS and SSH tunnels [36]. Users tracking and deanonymization via browser fingerprinting (by using JavaScript, timing and session identifiers) were proposed by Eckersley [37] and Károly [38]. In contrast to the aforementioned methods, our method is the first one to present the architecture and implementation of multi-layered and cross-layered systems to grade the authenticity of clients and connection peers. In addition, to the best of our knowledge, we are the first to propose a cross-view between passive and active traffic to detect anonymized or impersonated clients.

6   Conclusion

In the modern cyber-security arena, attackers and adversaries usually try to disguise their origin in order remain anonymized. As a defensive measure, it is important to identify which connections have initiated from anonymized or masked clients in order to block or investigate them further. In this work, we present AuthentIx, a system aimed at measuring the authenticity of a client in Internet traffic. Our system consists of a set of analytics engines which analyze traffic from passive or active sources. Authentix also includes a connection to different external databases, such as blacklisting/whitelisting of IPs and DNS, and address information and address reputation databases. We implement seven analytics engines and evaluate then against real anonymized clients. Our implementation includes passive and active traffic analytics, as well as cross-view between different protocol layers (e.g., request time interval ratio comparisons) and methods (OS active vs. passive fingerprinting). Our evaluation includes testing a real server with various clients, each of which employs a different type of anonymization method. The results shows that AuthentIx was able to detect all anonymized clients, and hence can be an efficient tool in real-time detection of suspicious connections. Furthermore, AuthentIx can be extended to cope with new type of anonymization and impersonation techniques used by attackers in the future.

References

1. Wheeler, D.A., Larsen, G.N.: Techniques for cyber attack attribution. Institute for Defense Analyses (2003)
2. Tangil-Rotaeche, D., Suarez, G., Palomar-González, E., Ribagorda-Garnacho, A., Ramos-Álvarez, B.: Anonymity in the service of attackers. Serb. Publ. InfoRev. Joins UPENET Netw. CEPIS Soc. J. Mag. **27**, (2010)
3. Hirt, A., Aycock, J.: Anonymous and malicious. In: 15th Virus Bulletin International Conference (2005)
4. Dingledine, R., Mathewson, N., Syverson, P.: Tor: the second-generation onion router (2004)
5. Fielding, R., Gettys, J., Frystyk, H., Berners-Lee, T.: RFC 2068: Hypertext Transfer Protocol —HTTP/1.1, January 1997. http://www.cis.ohio-state.edu/htbin/rfc/rfc2068.txt
6. https://kproxy.com/
7. http://anonymouse.org/
8. Leech, M., Ganis, M., Lee, Y., Kuris, R., Koblas, D., Jones, L.: SOCKS Protocol Version 5 (1996). https://www.ietf.org/rfc/rfc1928.txt
9. Robertson-Dunn, B.: Defcon Proxy Opens For Business (1999). http://mailman.anu.edu.au/pipermail/link/1999-June/039422.html
10. https://www.jmarshall.com/tools/cgiproxy/
11. https://www.anonymizer.com/
12. Hamzeh, K., Pall, G., Verthein, W., Taarud, J., Little, W., Zorn, G.: Point-to-point tunneling protocol (pptp) (1999). https://www.ietf.org/rfc/rfc2637.txt
13. Openvpn: Openvpn overview. https://openvpn.net/index.php/open-source/333-what-is-openvpn.html
14. Tech and Dev: Hiding Your IP Address Using SSH Tunneling Tutorial (2014). http://www.tech-and-dev.com/2014/04/hiding-your-ip-address-using-ssh-tunneling-tutorial.html
15. http://lcamtuf.coredump.cx/p0f3/

16. https://panopticlick.eff.org/
17. MaxMind Proxy Detection. https://www.maxmind.com/en/proxy-detection-service
18. http://www.ipvoid.com/
19. Blacklists, D.N.S.: http://www.dnsbl.info/
20. https://torstatus.blutmagie.de/
21. https://db-ip.com/db/
22. https://www.whois.net/
23. Koukis, D., Antonatos, S., Antoniades, D., Markatos, E.P., Trimintzios, P.: A generic anonymization framework for network traffic. In: 2006 IEEE International Conference on Communications (2006)
24. Goldschlag, D., Reed, M., Syverson, P.: Onion routing. Commun. ACM **42**, 39 (1999)
25. Reed, M.G., Syverso, P.F., Goldschlag, D.M.: Anonymous connections and onion routing. IEEE J. Sel. Areas Commun. **16**, 482 (1998)
26. Syverson, P.F., Tsudik, G., Reed, M., Landwehr, C.: Towards an analysis of onion routing security. In: Federrath, H. (ed.) Designing Privacy Enhancing Technologies. LNCS, vol. 2009, pp. 96–114. Springer, Heidelberg (2001). doi:10.1007/3-540-44702-4_6
27. Gabber, E., Gibbons, P.B., Kristol, D.M., Matias, Y., Mayer, A.: Consistent, yet anonymous, Web access with LPWA. Commun. ACM **42**, 42 (1999)
28. Freedman, M.J., Morris, R.: Tarzan: a peer-to-peer anonymizing network layer. In: Proceedings of the 9th ACM Conference on Computer and Communications Security (2002)
29. Langheinrich, M.: A privacy awareness system for ubiquitous computing environments. In: Borriello, G., Holmquist, L.E. (eds.) UbiComp 2002. LNCS, vol. 2498, pp. 237–245. Springer, Heidelberg (2002). doi:10.1007/3-540-45809-3_19
30. Schanzenbach, M.: Hiding from Big Brother. https://www.net.in.tum.de/fileadmin/TUM/NET/NET-2014-03-1.pdf#page=77
31. Savchenko, I.I., Gatsenko, O.Y.: Analytical review of methods of providing internet anonymity. Autom. Control Comput. Sci. **49**(8), 696–700 (2015)
32. Farah, T.: Traffic, Algorithms and Tools for Anonymization of the Internet. http://www2.ensc.sfu.ca/~ljilja/cnl/pdf/Thesis_tanjila_final.pdf
33. Coull, S.E., Wright, C.V., Keromytis, A.D., Monrose, F., Reiter, M.K.: Taming the devil: techniques for evaluating anonymized network data. In: Proceedings of Network and Distributed System Security Symposium (2008)
34. Sun, Y., Edmundson A., Vanbever, L., Li, O., Rexford, J., Chiang, M., Mittal, P.: RAPTOR. In: 24th USENIX Security Symposium (2015)
35. Manils, P., Abdelberri, C., Blond, S.L., Kaafar, M.A., Castelluccia, C., Legout, A., Dabbous, W.: Compromising Tor anonymity exploiting P2P information leakage. arXiv:1004.1461
36. Liming, L.: Traffic Monitoring and analysis for source identification (2010). http://scholarbank.nus.edu.sg/handle/10635/23750
37. Eckersley, P.: How unique is your web browser? In: Atallah, M.J., Hopper, N.J. (eds.) PETS 2010. LNCS, vol. 6205, pp. 1–18. Springer, Heidelberg (2010). doi:10.1007/978-3-642-14527-8_1
38. Boda, K., Földes, Á.M., Gulyás, G.G., Imre, S.: User tracking on the web via cross-browser fingerprinting. In: Laud, P. (ed.) NordSec 2011. LNCS, vol. 7161, pp. 31–46. Springer, Heidelberg (2012). doi:10.1007/978-3-642-29615-4_4

Statistical Network Anomaly Detection: An Experimental Study

Christian Callegari[1]([✉]), Stefano Giordano[2], and Michele Pagano[2]

[1] RaSS National Laboratory, CNIT, Pisa, Italy
christian.callegari@cnit.it
[2] Department of Information Engineering, University of Pisa, Pisa, Italy
{stefano.giordano,michele.pagano}@iet.unipi.it

Abstract. The number and impact of attack over the Internet have been continuously increasing in the last years, pushing the focus of many research activities into the development of effective techniques to promptly detect and identify anomalies in the network traffic. In this paper, we propose a performance comparison between two different histogram based anomaly detection methods, which use either the Euclidean distance or the entropy to measure the deviation from the normal behaviour. Such an analysis has been carried out taking into consideration different traffic features.

The experimental results, obtained testing our systems over the publicly available MAWILAb dataset, point out that both the applied method and the chosen descriptor strongly impact the detection performance.

1 Introduction

The ever increasing number of attacks over the Internet and the serious consequences that these can have in the citizens life have pushed the focus of many research activities into the design and development of effective tools to promptly detect and identify anomalies in the network traffic. As a result, many different approaches have been proposed in the last decade, but the ultimate solution is still far from being identified.

Among the different proposals, promising results are offered by the methods based on the estimation of the distribution of a given traffic feature (histogram based methods).

Nonetheless, even these anomaly detection systems are still affected by serious limitations (mainly in terms of missed detections and false alarms), either due to the intrinsic inability of the chosen method to deal with some kind of anomalies or to the low appropriateness of the chosen traffic feature to properly discriminate between normal and anomalous activities.

For this reason, in this paper, we propose an experimental study of two distinct approaches:

- one based on the computation of the Euclidean distance between the histograms of a given traffic descriptor computed in different time-bins (namely the current time-bin and a reference anomaly-free time-bin);

© Springer International Publishing AG 2016
R. Doss et al. (Eds.): FNSS 2016, CCIS 670, pp. 12–25, 2016.
DOI: 10.1007/978-3-319-48021-3_2

– one based on the variation of the entropy associated to the histograms of a given traffic descriptor computed in different time-bins.

Apart from simply comparing such methods, in this paper we have also investigated if their "relative" performance is constant when varying the considered traffic metric. In other words, we have verified if we can identify a method that outperform the other despite the chosen traffic descriptor. For this reason, focusing on volume anomalies as representative of a very widespread phenomenon, we have chosen to take into consideration two distinct traffic descriptors, namely the number of distinct flow destined to a given traffic aggregate and the quantity of bytes received by the same aggregate.

Interestingly, the experimental results, obtained testing our systems over the publicly available MAWILAb dataset, show that not only does the choice of both the statistical detection method and the considered traffic descriptor have a strong impact on the performance, but that the choice of the "best" detection method also depends on the considered traffic feature (as pointed out by the experimental results).

It is worth highlighting that for addressing the scalability issues, both the methods work on top of traffic aggregates (not traffic flow). Given the literature on the topic (see next section for more details), we have chosen to aggregate the traffic using probabilistic data structures (i.e., reversible sketches).

The rest of the paper is organised as follows: Sect. 2 gives a brief overview of the related works, and in Sect. 3 we provide a quick review of some background knowledge. Then in Sect. 4 we detail the proposed anomaly detection method. Hence, the used data-set for the experimental tests is described in Sect. 5, and the achieved performance is discussed in Sect. 6. Finally, Sect. 7 concludes the paper with some final remarks.

2 Related Work

Anomaly detection is a general framework including different analysis techniques, so it is not surprising that several works have been published in recent years, dealing with specific methods or providing a *general overview* of the different approaches (see, for instance, [1–3], which focuses on the features of network data and provides general guidelines for the design of IDSs). In the following, we only discuss the most relevant contributions closely related to this work.

Sketches, by themselves, cannot be considered as a detection method, but they are frequently used as a building block of several IDSs [4–8]. Indeed, random aggregation performed by sketches "efficiently" reduces the dimension of the data (wrt "classical" deterministic aggregations, such as according to input/output routers [9]); moreover, through the use of reversible sketches [10] it is possible to identify the flows responsible for the anomalies.

Regarding histogram based IDSs, in [11] the behavior of the monitored network during every time bin is characterized by means of histograms representing the distribution of the number of flows, packets or bytes over the values of a traffic feature. Anomalies are detected by comparing, through a distance function (namely, Euclidean distance, Manhattan distance, Mahalanobis distance, Kullback-Leibler divergence, and Jensen-Shannon divergence) the current histogram with a reference one, built during the training phase. In [12] the histogram cloning method is introduced: multiple randomized histograms are obtained through independent hash functions and the Kullback-Leibler divergence is used to detect anomalies.

Entropy has been applied to intrusion detection in different frameworks. For instance, in [13] fast Internet worms are detected taking into account the entropy contents (more precisely, the Kolmogorov complexity) of traffic parameters, such as IP addresses, while [14] focuses on network traffic running over TCP. In both cases an upper bound of Shannon entropy has been estimated through the use of different state-of-the-art compressors. Instead, in [15] Shannon entropy "summarizes" the distribution of specific traffic features to detect unusual traffic patterns.

3   Theoretical Background

In this section we present some theoretical background information, necessary to understand the proposed architecture. Note that we focus on the useful details only, referring the reader to the provided references for a complete description of the different topics.

3.1   Reversible Sketches

A sketch is a probabilistic data structure (a two-dimensional array) that can be used to summarize a data stream, by exploiting the properties of the hash functions [6]. Sketches differ in how they update hash buckets and use hashed data to derive estimates.

In more detail, a sketch is a two-dimensional $D \times W$ array $T_{D \times W}$, where each row d $(d = 0, \cdots, D - 1)$ is associated to a given hash function h_d. These functions give an output in the interval $(0, \cdots, W - 1)$ and these outputs are associated to the columns of the array. As an example, the element $T[d][j]$ is associated to the output value j of the hash function h_d.

When a new item (i_t, c_t), where i_t is the key (e.g., a destination IP address) and c_t is the weight (e.g., the number of received bytes), arrives, the sketch is updated as follows:

$$T[d][h_d(i_t)] \leftarrow T[d][h_d(i_t)] + c_t \tag{1}$$

and the update procedure is repeated for all the different hash functions.

Given the use of the hash functions, such data structures are not reversible, which makes impossible, after the detection, to identify the IP addresses responsible of an anomaly. To overcome such a limitation, in our system we have used an improved version of the sketch, that is the reversible sketch [10].

3.2   Entropy

The most basic concept in information theory is the entropy of a random variable (RV) X, often called Shannon entropy [16]. Roughly speaking, it is a measure of the uncertainty (or variability) associated with the RV.

In more detail, let $P = \{p_1, p_2, \ldots, p_L\}$ be the probability distribution of the discrete RV X, i.e.

$$0 \leq p_l \leq 1 \quad \text{and} \quad \sum_{l=1}^{L} p_l = 1$$

Then its Shannon entropy is defined as follows:

$$H(X) \;=\; -\sum_{l=1}^{L} p_l \log_2 p_l = \mathbb{E}\big[-\log_2 P(X)\big] \tag{2}$$

where $\mathbb{E}$ denotes the expectation operator, and is measured in bits (or shannon). Note that a change in the base of the logarithm just corresponds to a multiplication by a constant and a change in the unit of measure (nat for the natural logarithm and hartley (or ban) for the base 10 logarithm). In particular, when the natural algorithm is considered, (2) coincides with the well-known Boltzman–Gibbs entropy in statistical mechanics.

It is well-known that $0 \leq H(X) \leq \log_2 L$, where the infimum corresponds to the degenerate distribution (i.e., $p_l = \delta_{k-l}$ for some integer k with $1 \leq k \leq L$) and the supremum is attained in case of uniform distribution (i.e., $p_l = 1/L \; \forall l$).

3.3   Euclidean Distance

The Euclidean distance (or Euclidean metric) corresponds to the usual distance between two points in an Euclidean space (in $\mathbb{R}^2$ it is equivalent to the well-known Pythagorean theorem). It can be seen as a special case (for $p = 2$) of the Minkowski distance of order p

$$d_p(P, Q) \;=\; \left(\sum_{l=1}^{L} |p_l - q_l|^p\right)^{1/p}$$

We recall that for $p \geq 1$, the Minkowski distance is a metric (as a result of the Minkowski inequality); instead for $p < 1$ the triangle inequality does not hold (see, for instance, [17] for further details).

4   System Architecture

The proposed system takes as input traffic data over a predefined time-bin (in the following we assume to have N distinct time-bins), whose length can be arbitrarily set by the network administrator. Note that the duration of the time-bin is a compromise between the detection delay (the decision is taken at the end of the time bin) and the need of collecting enough data in order to build significant statistics. In more detail, the information associated to each time-bin is a list of keys i_t (e.g., the list of destination IP addresses) observed during that time-bin and the associated weights c_t (in our case, the number of bytes and flows for that IP address). Such information can be easily extracted from standard network traffic data, for instance parsing NetFlow traces by using the Flow-Tools [18].

The input data are processed to build the reversible sketch tables. In our case, each bucket will contain an histogram, representing the empirical distribution (estimated over L bins) of the weight values associated to all the keys that are mapped, by the corresponding hash function, in the given bucket. In this way, we have obtained T distinct sketches $T_{D \times W \times L}^t$, where $t \in [1, N]$ is the time-bin (in the experimental tests we have set $W = 512$, $D = 16$, and $L = 64$).

Then, the sketches are passed to the actual anomaly detection phase, where, for each bucket of the current sketch $T^t[d][w][\cdot]$, the system performs one of the following operations:

- entropy based method: the system computes the entropy associated to the current histogram and the difference between such a value and the entropy associated to the same bucket in the reference sketch (i.e., the last non-anomalous processed sketch);
- distance based method: the system computes the Euclidean distance between the current histogram and the histogram stored in the same bucket of the reference sketch.

Finally, such a value (either the entropy difference or the Euclidean distance) is compared with a threshold to decide if there is an anomaly or not. Note that, given the nature of the sketches, each traffic flow is part of D random aggregates and hence it will be checked D times to verify if any anomaly is present (indeed, an anomalous flow could be masked in a given traffic aggregate, while being detectable in another one).

Due to this fact, a voting algorithm is applied for each time-bin: the algorithm simply verifies if at least H (where H is a tunable parameter, with $H = D/2 + 1$ in our experiments) rows of the sketch contain at least one anomalous bucket. If so, the system reveals an anomaly and the responsible IP addresses are identified (by using the reversible sketch functionalities [10]).

5 MAWILab Dataset

The dataset used to evaluate our anomaly detection methods consists of packet traces from the MAWI (Measurement and Analysis on the WIDE Internet) archive (sample-points B and F), publicly available at [19]. Each trace in this database collects the traffic captured for 15 min in a specific day, since 2001 until nowadays, on a trans-Pacific link between Japan and the USA.

As in almost all existing databases, the key problem in testing the IDS performance is represented by a precise knowledge of the anomalies existing in the captured traffic. Such information is essential for building a proper ROC (Receiver

Operating Characteristic) curve and evaluating new approaches. Although also for the MAWI archive, an exact description of the attacks is not available, the dataset presents two important features that make it suitable for the performance evaluation procedure:

- unlike the widely-used DARPA dataset, the network is not emulated and the traffic mixture is representative of the current mixtures of network services and applications;
- in the framework of the successive project MAWILab [20], every traffic flow is classified by means of labels, which indicate the probability (according to well-known anomaly detection algorithms) that an anomaly is present. Since these labels are available together with the traces, they can be used as a common reference for testing a new IDS.

In more detail, the traces classification has been obtained combining the output of four anomaly detectors (based respectively on the Hough transform, the Gamma distribution, the Kullback-Leibler divergence and the Principal Component Analysis) [21]. As a result, the traffic is split into four categories:

- *anomalous*: traffic that is anomalous with high probability;
- *suspicious*: traffic that is probably anomalous, but not clearly identified by the MAWI classification methods;
- *notice*: non anomalous traffic, but that has been reported by at least one of the four anomaly detectors;
- *benign*: normal traffic.

The anomalies (*anomalous* and *suspicious* flows) are listed in an xml file for each trace, identifying them by means of traffic features as source and destination IP addresses, source port, destination port and transport protocol. Furthermore, some information about the kind of anomaly are also given:

- *attack*: anomalies representing a well known attack;
- *special*: anomalies involving well known ports;
- *unknown*: unknown kinds of anomalies.

Hence, the effectiveness of an IDS can be evaluated comparing the alarms generated by the new IDS with the labeled flows in the traffic traces, possibly referring to the three above-mentioned anomalous behaviors. Nevertheless, it is important to take into account the probabilistic nature of the MAWI classification in the interpretation of the achieved results.

6   Experimental Results

In this section we discuss the experimental results over the MAWILab dataset. The most widely used performance indicators are represented by the ROC curve and the Area under the Curve (AuC). Taking into account the MAWI labels, we consider as "false positives" the flows that are not labeled as "anomalous" or "suspicious" in the MAWI archive, but that are anomalous according to the tested IDS, so the false alarm probability P_{FA} is the ratio between the number of "false positive flows" and the number of flows that are neither "anomalous" nor "suspicious".

On the other hand, the false negative rate P_{FN} (note that the detection probability P_D can be obtained simply as $P_D = 1 - P_{FN}$) is the ratio between the number of false negatives and the number of "anomalous" flows. But, in this case P_{FN} depends on the actual interpretation of the MAWILab labels, and can be defined in several ways.

In more detail, as discussed in [22], the number of false negatives can be calculated as (the labels are used in the following figures to identifies the corresponding definitions of P_D):

– "all": the number of unrevealed flows labeled as "anomalous";
– "fn 2/3/4 detector": the number of unrevealed flows labeled as "anomalous" and detected at least by two/three/four of the four detectors used in MAWI classification;
– "fn attack": the number of unrevealed flows labeled as "anomalous" belonging to the "attack" category (known attacks);
– "fn attack special": the number of unrevealed flows labeled as "anomalous" belonging to the "attack" category or the "special" category (attacks involving well-known ports);

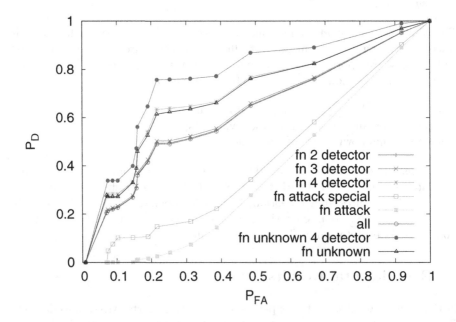

Fig. 1. ROC curves - entropy (flow)

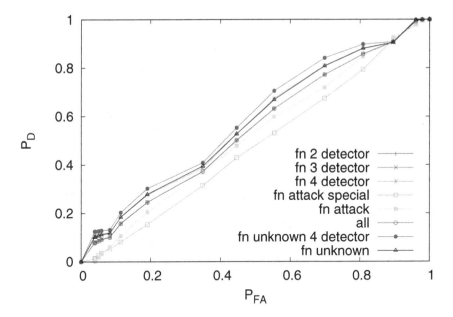

Fig. 2. ROC curves - Euclidean distance (flow)

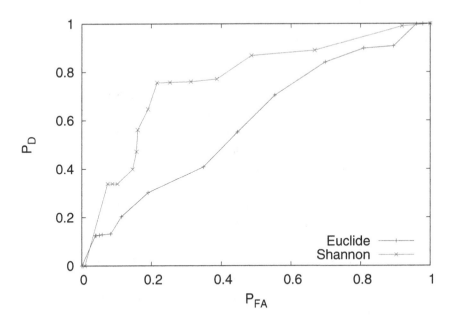

Fig. 3. ROC curves - comparison (flow)

- "fn unknown": the number of unrevealed flows labeled as "anomalous" belonging to the "unknown" category (unknown anomalous activities);
- "fn unknown 4 detector": the number of unrevealed flows labeled as "anomalous" belonging to the "unknown" category and detected by all the four detectors used in MAWI classification.

Given these definitions, in the following we discuss the results achieved by our system when taking into consideration, as traffic descriptors, either the number of flows with the same destination IP address (referred to as Flow in the following) or the quantity of traffic received by each IP address expressed in bytes (referred to as Byte in the following).

The first set of figures (namely Figs. 1, 2, and 3) refers to the Flow case. In more detail Fig. 1 shows the results achieved when using the entropy, for all the above mentioned definitions of P_{FN}. As it appears clearly, the offered performance strongly depends on the definition of P_{FN}, ranging from the completely unacceptable cases of "fn attack" and "fn attack special" to the very good case of "fn unknown 4 detector" (with a detection rate of about 80 % in correspondence of a false alarm rate less than 20 %). These results are very promising, taking into consideration that the usage of an anomaly detection system (normally in cascade to a misuse-based detection system) is conceived for detecting the "unknown" anomalies (given that the known attacks can be better detected by the other system).

Table 1. AuC (Flow)

Method	Label	AuC
Euclidean distance	All	0.546382
Euclidean distance	fn 2 detector	0.546917
Euclidean distance	fn 3 detector	0.546582
Euclidean distance	fn 4 detector	0.570564
Euclidean distance	fn attack	0.520335
Euclidean distance	fn attack special	0.481449
Euclidean distance	fn unknown	0.57054
Euclidean distance	fn unknown 4 detector	0.590988
Entropy	All	0.61949
Entropy	fn 2 detector	0.621851
Entropy	fn 3 detector	0.627007
Entropy	fn 4 detector	0.699259
Entropy	fn attack	0.362535
Entropy	fn attack special	0.418059
Entropy	fn unknown	0.693165
Entropy	fn unknown 4 detector	0.768803

Figure 2 presents the same analysis when using the Euclidean distance, showing that such a method is far from providing good results, having for all of the plots a behaviour very close to the diagonal.

A more precise comparison between the two methods is shown in Table 1 and in Fig. 3, where we respectively present the AuC obtained by the two methods when varying the definition of P_{FN} and the ROC achieved in the "fn unknown 4 detector" case. Hence, we can easily conclude that the entropy method definitely offers better performance than Euclidean distance when using Byte as traffic descriptor.

A completely analogous performance analysis is presented in the subsequent figures and table, where we show the ROCs and the AUC values for the two systems, obtained when using Byte as traffic descriptor. In this case it is very interesting to make two observations:

– the offered performance is, in any case, very far from those related to the use of Flow as traffic descriptor (see Figs. 4 and 5, where "almost unacceptable" ROC are shown), demonstrating how the choice of the correct traffic descriptor is crucial in anomaly detection;
– contrarily to the Flow case, better performance is offered by the Euclidean distance (see Fig. 6) , demonstrating how the anomaly detection method must be properly chosen (also taking into account the used traffic descriptor).

Finally, Table 2 presents all the values of the AuC for the Byte case, confirming the results shown in Figs. 4 and 5.

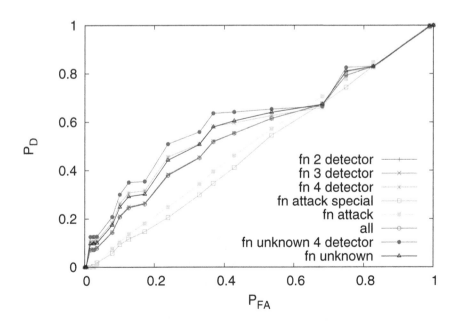

Fig. 4. ROC curves - Euclidean distance (byte)

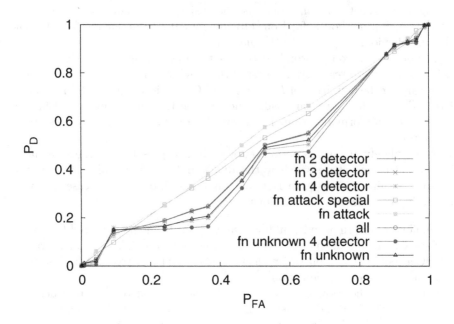

Fig. 5. ROC curves - Entropy (byte)

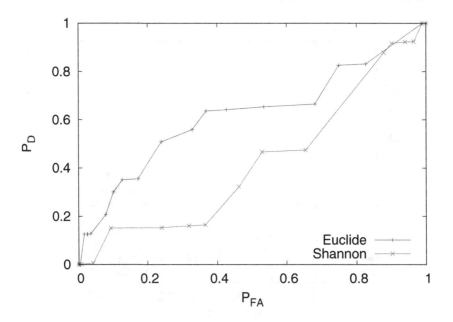

Fig. 6. ROC curves - comparison (byte)

Table 2. AuC (Byte)

Method	Label	AuC
Euclidean distance	All	0.566218
Euclidean distance	fn 2 detector	0.566777
Euclidean distance	fn 3 detector	0.567148
Euclidean distance	fn 4 detector	0.593179
Euclidean distance	fn attack	0.517885
Euclidean distance	fn attack special	0.49092
Euclidean distance	fn unknown	0.59376
Euclidean distance	fn unknown 4 detector	0.619295
Entropy	All	0.455644
Entropy	fn 2 detector	0.455011
Entropy	fn 3 detector	0.453766
Entropy	fn 4 detector	0.436463
Entropy	fn attack	0.515499
Entropy	fn attack special	0.496697
Entropy	fn unknown	0.440622
Entropy	fn unknown 4 detector	0.421189

7   Conclusion

In this paper we have proposed an experimental comparison between tow different histogram based anomaly detection methods. Moreover, the impact of the considered traffic descriptor on the achieved performance has been investigated.

The experimental results, obtained testing our systems over the publicly available MAWILAb dataset, have clearly demonstrated that

- the choice of the correct traffic descriptor is crucial in anomaly detection;
- the anomaly detection method must be properly defined (also taking into account the used traffic descriptor).

These results show that the deployment of an anomaly detection tool requires a fine tuning of the system, also based on a good knowledge of the considered network scenario and traffic.

Acknowledgment. This work was partially supported by Multitech SeCurity system for intercOnnected space control groUnd staTions (SCOUT), a FP7 EU project.

References

1. Thottan, M., Liu, G., Ji, C.: Anomaly detection approaches for communication networks. In: Cormode, G., Thottan, M., Sammes, A.J. (eds.) Algorithms for Next Generation Networks. Computer Communications and Networks, pp. 239–261. Springer, London (2010)
2. Ahmed, M., Naser Mahmood, A., Hu, J.: A survey of network anomaly detection techniques. J. Netw. Comput. Appl. **60**(C), 19–31 (2016)
3. Callegari, C., Coluccia, A., D'Alconzo, A., Ellens, W., Giordano, S., Mandjes, M., Pagano, M., Pepe, T., Ricciato, F., Zuraniewski, P.: A methodological overview on anomaly detection. In: Matijasevic, M., Callegari, C., Biersack, E. (eds.) Data Traffic Monitoring and Analysis. LNCS, vol. 7754, pp. 148–183. Springer, Berlin (2013)
4. Subhabrata, B.K., Krishnamurthy, E., Sen, S., Zhang, Y., Chen, Y.: Sketch-based change detection: methods, evaluation, and applications. In. Internet Measurement Conference, pp. 234–247(2003)
5. Borgnat, P., Dewaele, G., Fukuda, K., Abry, P., Cho, K.: Seven years and one day: sketching the evolution of internet traffic. In: INFOCOM, April 2009
6. Cormode, G., Muthukrishnan, S.: An improved data stream summary: the count-min sketch and its applications. J. Algorithms **55**(1), 58–75 (2005)
7. Lakhina, A., Crovella, M., Diot, C.: Mining anomalies using traffic feature. In: ACM SIGCOMM (2005)
8. Salem, O., Vaton, S., Gravey, A.: A scalable, efficient and informative approach for anomaly-based intrusion detection systems: theory and practice. Int. J. Netw. Manag. **20**, 271–293 (2010)
9. Callegari, C., Gazzarrini, L., Giordano, S., Pagano, M., Pepe, T.: When randomness improves the anomaly detection performance. In: Proceedings of 3rd International Symposium on Applied Sciences in Biomedical and Communication Technologies (ISABEL) (2010)
10. Schweller, R., Gupta, A., Parsons, E., Chen, Y.: Reversible sketches for efficient and accurate change detection over network data streams. In: Proceedings of the 4th ACM SIGCOMM Conference on Internet Measurement. IMC 2004, pp. 207–212. ACM, New York (2004)
11. Kind, A., Stoecklin, M.P., Dimitropoulos, X.: Histogram-based traffic anomaly detection. IEEE Trans. Netw. Serv. Manag. **6**(2), 110–121 (2009)
12. Brauckhoff, D., Dimitropoulos, X., Wagner, A., Salamatian, K.: Anomaly extraction in backbone networks using association rules. IEEE/ACM Trans. Netw. **20**(6), 1788–1799 (2012)
13. Wagner, A., Plattner, B.: Entropy based worm and anomaly detection in fast IP networks. In: 14th IEEE International Workshops on Enabling Technologies: Infrastructure for Collaborative Enterprise (WETICE 2005), pp. 172–177, June 2005
14. Callegari, C., Giordano, S., Pagano, M.: On the use of compression algorithms for network anomaly detection. In: 2009 IEEE International Conference on Communications, pp. 1–5, June 2009
15. Lakhina, A.: Diagnosing network-wide traffic anomalies. In. ACM SIGCOMM, pp. 219–230 (2004)
16. Shannon, C.E., Weaver, W.: The Mathematical Theory of Communication. University of Illinois Press, Champaign (1949)

17. Kolmogorov, A., Fomin, S.: Elements of the Theory of Functions and Functional Analysis. Number v. 1 in Dover Books on Mathematics. Dover (1999)
18. Flow-Tools Home Page. http://www.ietf.org/rfc/rfc3954.txt
19. MAWI Working Group Traffic Archive. http://mawi.wide.ad.jp/mawi/. Accessed Nov 2011
20. MAWILab. http://www.fukuda-lab.org/mawilab/ Accessed Nov 2011
21. Fontugne, R., Borgnat, P., Abry, P., Fukuda, K.: MAWILab: combining diverse anomaly detectors for automated anomaly labeling and performance benchmarking. In: ACM CoNEXT (2010)
22. Callegari, C., Casella, A., Giordano, S., Pagano, M., Pepe, T.: Sketch-based multidimensional IDS: a new approach for network anomaly detection. In: IEEE Conference on Communications and Network Security, CNS 2013, National Harbor, MD, USA, 14–16 October 2013, pp. 350–358 (2013)

Power Usage Efficiency with a Modular Routing Protocol

Yoshihiro Nozaki[1]([✉]), Nirmala Shenoy[1], and Aparna Gupta[2]

[1] Rochester Institute of Technology, Rochester, NY, USA
yoshihihro.nozaki@fireeye.com, nxsvks@rit.edu
[2] Rensselaer Polytechnic Institute, Troy, NY, USA
guptaa@rpi.edu

Abstract. Recent years have seen major efforts to contain the environmental footprint of the Internet. The last decade has witnessed revolutionary research to address some of the challenges faced in the Internet. This article describes an investigative framework for determining the energy savings incurred with new routing protocols and routers. The framework is applied to a real ISP network - the AT&T ISP network in the United States. It describes techniques to collect statistics from such large networks and analyze them. The statistics are then used to study the energy consumption in the ISP network both with routers running the current routing protocol, which is Open Shortest Path First (OSPF) and also with routers running the new protocol. The cost models and energy savings studies applied to large ISP networks as presented in this article is the first of its kind. As evidenced from this study significant energy benefits and cost savings thereof can be realized with the proposed modular routing protocol.

Keywords: Modular routing architecture · Framework for energy cost studies · Transition costs and benefits · Operational complexity

1 Introduction

The Internet delivers millions of packets of information between networks across the globe to sustain the communication demands of organizations that depend on the Internet for their day-to-day operations. The massive and ubiquitous spread of the Internet in turn demands high computing devices to run the highly complex routing operations to deliver packets. Routers have grown in complexity over the years and consume a high amount of energy both for their operations and cooling. The carbon footprint of the Internet is also showing growth trends that demand serious consideration [20,23].

Energy consumption by the Information and Communication Technology (ICT) and energy-efficient networking are rapidly gaining the attention of the communications research community. This is also of major concern to the Internet Service Providers (ISP) and *Telecommunications* operators as their infrastructures are continuously growing. Networking devices such as routers

© Springer International Publishing AG 2016
R. Doss et al. (Eds.): FNSS 2016, CCIS 670, pp. 26–46, 2016.
DOI: 10.1007/978-3-319-48021-3_3

Table 1. Annual energy consumption by a few major telecom operators and estimated electricity cost. Source: [13, 29]

ISP	Energy consumption (2009)	Cost ($60/MWh)
AT&T	11.07×10^6 MWh	$664.2M
Verizon	10.27×10^6 MWh	$616.2M
NTT	2.75×10^6 MWh	$165.0M
China Mobile	10.62×10^6 MWh	$637.2M
France Telecom	4.38×10^6 MWh	$262.8M
Deutsche Telecom	7.91×10^6 MWh	$474.6M

are required to be powered up 24/7 and a number of these devices owned by the ISPs are high computing complex devices that have a high-energy consumption. Energy expenses are increasing. Table 1 shows annual energy consumption by some major ISPs and the estimated electricity costs, in millions of dollars. The increasing trend of the costs has been confirmed by reports from the industry [13]. Reduction in power consumption is of high priority to the ISPs and they are *seeking efficient protocols, architectural solutions, and innovative equipment that will consume less energy and produce sufficient performance* [17,21].

Power consumption in a router can be attributed primarily to the components required to run the routing protocol such as the forwarding engine, the I/O operations and control operations which account for 51.5 % of the energy consumed by a router [9]. Open Shortest Path First (OSPF) and Border Gateway Protocol (BGP) are the two main routing protocols used in the Internet. Both OSPF and BGP increase in their operational complexity as the number of networks increase. This is because the routing tables populated by both protocols have a direct dependency on the network numbers. Thus complexities in router operations and fast-memory technology (to quickly access routing databases) have been growing to sustain the growth in the Internet. Statistics show that the power dissipation in routing equipment is increasing at twice the rate of improvements in power consumption due to advances in hardware technologies [21] and the carbon footprint of the Internet now probably exceeds that of air travel by as much as a factor of two[1] [20].

Decoupling the routing table size dependency on the numbers of networks, should have a positive impact by reducing the complexity of routing operations. The algorithms underlying routing protocols, which control the routing operations should be investigated for this purpose. The link state routing algorithm of OSPF and path vector algorithm used by BGP do not allow for decoupling between routing table sizes and network numbers.

Research organizations all over the world are encouraging novel and clean slate revolutionary initiatives to solve some of Internet's primary concerns such

---

[1] The carbon footprint is not only due to routers but the excessive use of all types of computing devices on the Internet to support the services that are being offered.

as security, addressing and routing scalability among others [4,6,8]; As part of an NSF funded project [5], the authors of this article investigated the use of structures in networks to develop a modular routing architecture. A modular routing protocol was designed and developed on this modular routing architecture. The dependency of the routing table sizes on the network numbers was decoupled. Before adopting technologies that are the outcomes of revolutionary research, it is important to study the cost vs benefits of such an adoption.

As part of the cost versus benefit study, the energy benefits of adopting the modular routing protocol are presented in the article. The study follows the step-by-step approach outlined. (1) Adopt a real ISP network in the United States (US) namely the AT&T ISP network. (2) Derive the details of the AT&T ISP network from the Rocketfuel database [32] with information on the number of POPs, routers and links. (3) Introduce a technique to determine the number of backbone, distribution and access routers in the ISP network. (4) Determine the routing protocol operational complexity and the memory usage needs both for OSPF and the proposed modular routing protocol called the Tiered Routing Protocol (TRP) as applied to the AT&T ISP network. (5) Use the protocol operational complexity in OSPF and TRP to calculate a complexity index to assess the numbers of TRP backbone, distribution and access routers as compared to OSPF backbone, distribution and access routers. (6) Use these indices to determine the power consumption in routers that will run TRP. (7) Compute the difference in power consumption before and after all OSPF routers are potentially replaced with TRP routers in the AT&T ISP network in the US. (8) Lastly, a power usage case study of a transition scenario to replace existing OSPF routers with TRP routers using a stage-by-stage approach is presented.

Section 2 discusses some related work to reduce power consumption in the Internet. Some of the specification of the modular routing architecture and the operational details of the routing protocol are available in [24–26]. However, for quick reference and easy reading, Sect. 3 describes briefly the modular routing architecture and the protocol. Section 4 focuses on the power usage analysis based on statistics collected for the AT&T ISP network in the US, typical router models used in ISP networks along with their power consumption, and market prices. Results from our studies that provided information on the routing table sizes and control overheads when running OSPF and TRP in typical network topologies established over Emulab are presented and used in the power usage calculations. Section 5 is devoted to the transition case study and power savings calculations during transition. Section 6 summarizes this work with some conclusions.

2   Related Work

Recent years have seen major efforts to contain the environmental footprint of the Internet. *The recommended paradigm is produce adequate performance with minimum energy costs.* This defines a fundamental change in the way communications research is to be conducted namely sustainability should be an inherent part of network research [28].

The related work in this topic area can be categorized into (i) enhancement to routing operations through optimal paths, sleep states or reduced packet handling complexity in a session and (ii) improvement to hardware and software to contain the power consumption in routing equipment and fast-memory design to access routing databases quickly.

A collaborative approach between ASs to use lower power paths is proposed in [3]. In [10], selectively connecting end devices and in [18] techniques to put routers to sleep are proposed. An approach called session-flow based packet forwarding is discussed in [31] to reduce packet processing.

In terms of hardware design, in [15] larger and sophisticated architectures are being explored. Purpose built silicon is considered in [16], while costlier and efficient fast memory such as Ternary Content Addressable Memory (TCAM) are discussed in [33]. The authors in [24] considered fast router forwarding engines.

In this article, a framework to calculate energy savings incurred, when new and novel routing techniques are to be deployed, which require new routers is presented. The framework has been applied to a large ISP network and comparative energy calculations provided for the new routing protocol and router versus the current routing protocol and routers used in the ISP networks. Routing techniques that can significantly reduce energy consumption in a router are discussed and their energy consumption calculated and compared with current routing techniques.

3   A Modular Routing Architecture

Figure 1 shows the ISP tier structure existing in the Internet today. The Points of Presence (POPs) form the backbone of an ISP network. A POP comprises of several router types, such as Backbone routers (BB), Distribution Routers (DR) and Access Routers (AR). BB routers primarily connect to the BB routers in other POPs. The DRs connect the ARs to the BB routers and ARs connect to customer or stub networks. The BB routers belong to tier 1, the DRs to tier 2 and the ARs to tier 3 as noted on the right side of Fig. 1. Few details of the routing protocol and the derivation of tiered routing addresses (TRA) used by the routing protocol are now described for a sample POP network shown in Fig. 2.

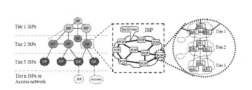

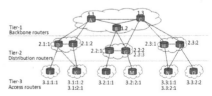

Fig. 1. ISP network tier structure   **Fig. 2.** Tiered addresses

The 3-tier structure in Fig. 2 can be noticed in any POP. A similar 3-tier structure consisting of backbone or core routers, distribution routers and access routers can also be noticed in any stub network.

3.1 Tiered Routing Addresses

The TRP uses the Tiered Routing Addresses (TRA) to perform packet forwarding and routing to distant networks/routers. The TRA in turn derives its format from the properties of the tier structures. A TRA is given by

$$TRA = TV.TreeAddress \tag{1}$$

where TV is the *Tier Value* and shows the tier in which the router or network resides, and the *TreeAddress* (TA) explicitly shows the parent-child relationships among connected routers.

The .(dot) notation separates the TV from the TA. Address allocation starts from the routers at tier 1. Routers A, B, and C at tier 1 are allocated TRAs 1.1, 1.2 and 1.3 respectively. Note that TRA assignment in TRP is *for a router*, and *not for each interface* in the router. This has advantages of reduced number of addresses, reduced routing table size, and ease in addition and removal of routers in the network. The TA of routers at tier 2 are allocated based on the TAs of their directly connected parent routers. Due to the parent-child relationship between Routers A and D, Router D's TA is derived by taking Router A's TA and appending a unique identifier for Router D. Hence, TRA of Router D following the format TV.TA is given by 2.1:1, where the two fields in the TA are separated by : (colon). Likewise, Router E gets a unique identifier 2 from Router A and its TRA is 2.1:2. A link between routers, which share a common parent, is called a trunk-link. Links between Routers D-E, F-G, and H-I are trunk-links, and are represented with dotted lines in Fig. 2.

Routing tables in the TRP routers are populated from 'hello' messages received from neighbor routers that carry their TRAs. The port of reception of the 'hello' messages is recorded as the port to access that neighbor. Typical routing tables populated by TRP at two Routers F and G (in Fig. 2) are given in Tables 2 and 3.

Packet Forwarding. Due to the properties of the tier structure and the explicit router parent-child relationships in the TRAs, routing a packet from a source

Table 2. Routing tables of Router F from Fig. 2

Router F with TRA {2.2:1}					
Uplink		Down		Trunk	
Port	Dest	Port	Dest	Port	Dest
1	1.2	3	3.2:2:1	2	2.2:2
					2.3:3

Table 3. Routing tables of Router G from Fig. 2

Router G with TRAs {2.2:2, 2.3:3}					
Uplink		Down		Trunk	
Port	Dest	Port	Dest	Port	Dest
1	1.2	3	3.2:2:1	4	2.2:1
2	1.3				

node to a destination node requires comparison of the tier values and a string comparison of the TAs in the source node TRA and destination node TRA to determine the packet forwarding direction. Details of packet forwarding algorithm are available in [25].

4   Power Usage Analysis

There are several steps towards determining the power usage costs by current OSPF routers and the potential TRP routers. As this is an applied study, the first step involved adopting a real ISP network namely the AT&T ISP network in the US and collecting information about its topology, the POPs and their numbers and locations, the number of links, and the numbers of backbone, distribution and access routers in each POP. To determine the power consumption in the different routers, typical models of routers used in the backbone, distribution and access routers and their power consumption and market price statistics were collected. The power consumption statistics included a breakdown in the power consumed by the different components such as forwarding engine, control operations and I/O memory operation among others. To determine power consumption by TRP routers, the power consumed by current OSPF routers, router operational complexity ratio of OSPF to TRP, and a ratio of memory usage by OSPF and TRP was calculated and used.

4.1   Topology and Router Statistics: The AT&T ISP Network

The topological, POP and router information of the AT&T ISP network were abstracted from the Rocketfuel database [32]. It was found that the AT&T ISP network had a total of 110 POPs, 11,403 routers and 13,689 links. The BB routers are high performance computing devices and cost more. Second in complexity and costs are the DRs. Hence, besides the POP information, we also abstracted the information pertaining to the numbers of BB, DR and AR routers in the AT&T ISP network. The BB routers in a POP were identified by their links to other POPs. The edge routers in the POP were identified as ARs. Routers connecting BB routers and ARs were then identified as the DR routers. This was the classification method adopted in this study[2]. Based on these classifications

---

[2] The methodology to identify BB, DR and AR routers and POPs was confirmed by Level 3 another ISP in the US.

Table 4. Breakdown of power consumption by a router. Source: [12]

Parts	% of Total power
Supply loss and blowers	35.0
Forwarding engine	33.5
Switching Fabric	10.0
Control plane	11.0
I/O	7.0
Buffers	3.5
Total	100

there were 389 BB routers, 6,395 DR routers, and 4,619 AR routers within the AT&T ISP network in the US [24].

Power Usage (PU) studies were conducted by considering the power consumed by the routing operations (software) and the hardware equipment separately. Hardware power usage is the wall-socket power and is constant for a given type of router. Software power usage depends on the CPU and memory usage in a router, which depends on the complexity of the routing operations. Besides, the power usage also changes with the type of router *i.e.* BB, DR or AR router. Furthermore the power consumed by a BB router in a TRP deployment will be different from the power consumed by a BB router in an OSPF deployment, as the operational complexity and memory needs of a TRP router will be different from that of an OSPF router. Same is true for the DR and AR routers. The studies thus differentiate not only the power consumption by the type of router, but also by the routing protocol running in the router.

From [12], we have Table 4 that shows the breakdown in power usage in a router. From this table it is clear that 33.5 % of power consumption in a router is due to the forwarding engine that implements the routing operation and performs forwarding. 7.0 % of the power is consumed by the memory storage need for routing tables and 11.0 % is consumed for handling the control operations due to routing control messages. Thus, to support routing operations alone 44.5 % of the power is consumed.

The next couple of sections are devoted to analyzing the operational complexity and memory needs for routers running OSPF as compared to routers running TRP.

4.2   Router Operational Complexity

To study power usage efficiency of an OSPF router vs a TRP router it is important to determine the operational complexity of the OSPF router vs the TRP router. Three factors contributing to the router operational complexity were considered in this study. They are (1) routing tables sizes (2) method of updating and maintaining the routing tables and (3) the control overhead generated in the event of link/node failures. In this sub-section, we look at statistics collected for routing table sizes and control overhead both for OSPF and TRP.

Routing Tables and Control Overhead. To determine the routing table sizes and the control overhead in the event of a link failure in the AT&T ISP network in the US, a two-stage approach is required. In stage 1, the routing table sizes and control overhead in networks running OSPF were determined by running a free OSPF software from QUAGGA [7] in real network topologies of up to 50 nodes (routers) set up in the Emulab test bed [2]. On the same topology the TRP software code written by us in C language (to run on Linux OS) was also run and the routing table sizes and control head in the event of a single failure was recorded. Details about this study is available in [25].

In stage 2, custom simulation software was written to calculate the routing table sizes and control overheads for any given ISP network topology, when running OSPF and TRP. This simulation software was validated with the data collected from real network topologies of up to 50 nodes from the Emulab testbed. The simulation software was then run for the AT&T ISP network topology collected from Rocketfuel database.

For the AT&T ISP network running OSPF in the US, the largest routing table size was determined to be 13,689 and around 300 routers had routing tables of this size. For routers running TRP the largest routing table size was 68, very few routers recorded this value. The routing table sizes so determined for OSPF and TRP are plotted in Fig. 3, where the x axis plots the incremental router number as the routing tables were computed for each router in the AT&T network. The plots are truncated to a little below 3,000 routers, because the routing tables recorded after this value are very low.

The control traffic generated and transmitted by a router in the event of link failures also contributes to the power consumption in a router. The overhead generated by OSPF and TRP to update in the event of a single link failure in the AT&T network is shown in Fig. 4. The x axis shows an incremental count in

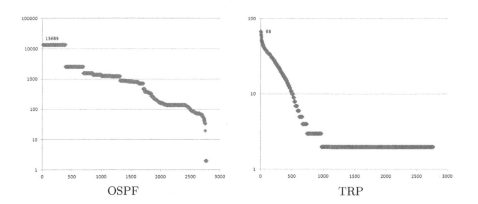

Fig. 3. Routing table sizes for OSPF and TRP in the AT&T network, USA

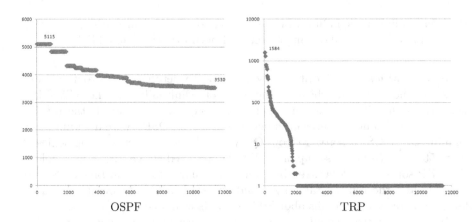

OSPF   TRP

Fig. 4. Number of update packets generated by OSPF and TRP for a single link failure in the AT&T network, USA

router numbers as the control overhead generated by each router was computed. This time the computation was stopped approximately at 12,000 routers. In the case of OSPF, the maximum number of update packets generated by a single failure is 5,115 and the minimum is 3,530. With TRP the maximum number of update packets is 1,584 and the minimum is 1. Average number of update packets generated by OSPF and TRP for a single failure is thus 4,003.66 and 25.64 respectively.

Routing Tables, Memory Usage and Energy Savings. Recent software routers and the older hardware router often use Static Random Access Memory (SRAM) or Dynamic Random Access Memory (DRAM) to store routing table entries. Longest prefix matching method for IP prefix lookup is used. The time complexity for the look up in this case is $O(\log(w))$, where w is the number of bits in the target IP address prefix [11]. Modern hardware routers however use Ternary Content Addressable Memory (TCAM), which allows parallel lookup with time complexity of $O(1)$. However, TCAMs also face the following limitations:

- *High power consumption*: TCAM consumes 12~15 Watts per chip (18MB) is and also proportional to the number of bits enabled in the TCAM during the search operation [30]. 4 to 8 TCAM chips are often used in a router [27].
- *Limited capacity*: Low cell density compared to SRAM cell that consist of 6 transistors, but TCAM has 16 transistors on a cell [34].
- *Cost*: TCAMs are expensive.

Based on the plots for routing table sizes provided in Fig. 3, TRP routing table sizes are significantly lower than OSPF. Therefore, the number of memory chips used in a TRP router will be low compared to a router running OSPF.

Because of the small size of the routing table size a TRP router can also implement hash table lookup which has $O(1)$ time complexity. In addition, TRA addresses (maximum length of 12 bits) [26] are of reduced length as compared to IPv4/v6, which will lead to reduced power consumption in a memory chip.

In this study, we assume the use of SRAM/DRAM to reduce the higher power consumption by TCAM and to avail the higher cell density and lower cost of SRAMs. This information is used in Subsect. 4.4 below to calculate the memory needs to store routing tables and hence the energy consumption.

Routing Protocol Operational Complexity. The process of populating and updating routing tables is a big factor in the determination of routing protocol's complexity. OSPF routing protocol uses the Dijkstra algorithm to find the shortest path to every destination network. The complexity of OSPF is given by $O(r \times \log(r))$, where r is number of routers in the OSPF routing domain [22]. TRP does not perform any path calculations because entries in a TRP routing table are the addresses of direct neighbors. The protocol operational complexity of TRP is thus $O(1)$ if hash tables were used. Without hash tables, the complexity will be $O(n)$, where n is the number of direct neighbors. In this study, the TRP operational complexity $O(n)$ was used.

4.3   Router Complexity Computation

The above statistics reveals the reduction in operational complexity that can be achieved with TRP routers as compared to OSPF routers. Accordingly if backbone routers running OSPF were replaced with TRP routers the number of routers requiring the operational complexity of a current backbone router would be lower. That is, due to the reduced operational complexity with TRP, it would be possible to substitute the highly complex BB routers used in the POP running OSPF with routers of operational complexity equivalent to a DR router running TRP. Applying the same argument a DR router used in a current POP running OSPF can be substituted with a TRP router with the capability of the current AR router. This requires a substitution criteria based on the routing operational complexity of an OSPF router vs TRP router.

The Substitution Criteria. The ratio of router operational complexity of an OSPF vs TRP router will be used to determine the substitution criteria.

$$C_{ratio} = \frac{O(r \times \log(r))}{O(n)} \tag{2}$$

Generalizing this to a router x that has R routers in its domain in the case of OSPF and N direct neighbors in the case of TRP, we get

$$C_{ratio}(x) = \frac{O(R(x) \times \log(R(x)))}{O(N(x))} \tag{3}$$

Table 5. Models of routers and prices [1]

Types	Models	Prices
Access Router (AR)	Cisco 7603	$10K
Distribution Router (DR)	Cisco 12816	$22K
Backbone Router (BB)	Cisco CSR-1	$100K

Thus a BB router running OSPF can be substituted by a TRP router of DR capacity (to perform the backbone operations) and a DR router running OSPF can be substituted by a TRP router of AR capacity (to perform the distribution operations) if the operational complexity ratio *i.e.* $C_{ratio}(x)$ of OSPF to TRP at router x attains a certain threshold value. To determine this threshold value we need another index that provides a direct inference of the router complexity. For this purpose we use the market prices of the BB, DR and AR routers. The market price of a router is proportional to the operational complexity of a router[3].

Threshold Index for Substitution. To determine the threshold complexity index, we obtained the details on router models used in BB, DR and AR routers from [9]. We then obtained the prices for these router from [1]. The router models and the price details are provided in Table 5.

From Table 5, the price ratio:

$$[BB_{price} : DR_{price} : AR_{price}] = [10.00 : 2.20 : 1.00] \tag{4}$$

Equation (4) can be used as a ratio that provides the proportional operational complexity between a BB, DR and AR router. Thus a BB router is 4.55 (10/2.2) times more complex in its operations than a DR router and a DR router is 2.2 times as complex in operations as an AR router.

Hence if $C_{ratio}(x)$ from Eq. (3) exceeds 4.55 where x is a BB router, then it can be substituted by a TRP router of capacity equivalent to a current DR router.

And if $C_{ratio}(x)$ from Eq. (3) exceeds 2.20 and x is a DR router, then the DR router can be substituted by a TRP router of capacity equivalent to an AR router.

The Substitution Process. Let BB_{OSPF} be the number of BB routers, DR_{OSPF} be the number of DR routers, and AR_{OSPF} be the number of AR routers in the current POP deployments in an ISPs network. Let BB_{TRP} be the number of BB routers, DR_{TRP} be the number of DR routers, and AR_{TRP} be the number of AR routers if instead of OSPF we were running TRP in the ISP network. Given that with TRP deployment we will be substituting some BB

---

[3] We are assuming that the router price is directly proportional to the complexity of the router.

routers with DR routers and some DR routers with AR routers the following equations hold.

$$BB_{TRP} = BB_{OSPF} - \Delta BB_{OSPF} \tag{5}$$

where ΔBB_{OSPF} is the number of BB_{OSPF} routers that can be substituted by a DR_{TRP} router. Hence routers of capacity as required in the backbone routers will now be lower if TRP routers were deployed. Or in other words, a TRP router of DR capacity can handle the backbone routing operations. Hence, Eq. (6), which is the calculation of DR routers with TRP deployment, will be given by

$$DR_{TRP} = DR_{OSPF} - \Delta DR_{OSPF} + \Delta BB_{OSPF} \tag{6}$$

here ΔDR_{OSPF} is the number of DR_{OSPF} routers that can be replaced with AR routers because of the reduced operational complexity needs with TRP.

Equation (7) gives the total number of AR routers with TRP deployment has increased because now the number of routers of AR capacity is more as some of the operations carried out by DR routers earlier can now be carried out by the AR routers, if TRP were deployed.

$$AR_{TRP} = AR_{OSPF} + \Delta DR_{OSPF} \tag{7}$$

Router Number Statistics - OSPF vs. TRP. Figure 5 shows the distribution of BB, DR, and AR router in each POP (total 110 POPs) in the AT&T network in the US as determined from the Rocketfuel database. For example, Chicago POP has a total of 1,010 routers of which there are 26 BB routers, 398 DRs, and 586 ARs. Using Eqs. (3) to (6) and after substituting the current OSPF BB, DR and AR routers with TRP BB, DR and AR routers, Fig. 6 is the plot of each type of router in each POP after the transition from OSPF to TRP.

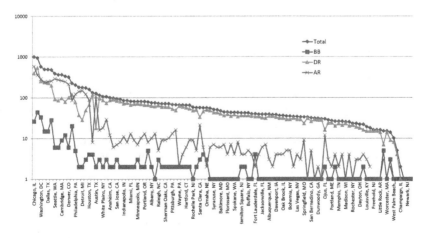

Fig. 5. Distribution of BB, DR, and AR routers in each POP (AT&T)

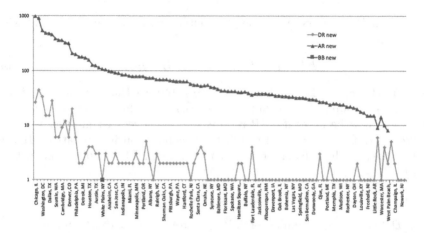

Fig. 6. Distribution of BB, DR, and AR routers in each POP after transition (AT&T)

Table 6. Total number of routers before and after transition

	BB	DR	AR	Total
Before	389	6395	4619	11403
After	1	399	11003	11403

Table 6 records the total number of routers of each capacity before and after the transition. From Table 6, it is to be noted that only 1 router will be needed that has the processing capacity of a current BB router. Instead of 6,395 routers that have the processing capacity required by current DRs, we will need only 399 of this capacity routers after the transition. The number of the low capacity router namely the ARs has increased. The numbers of BB, DR, and AR routers are different before and after the transition in the case of deployment with TRP routers and is a significant contributor to both energy and cost savings.

4.4 Router Power Usage Cost

To calculate the power consumption by the TRP routers we followed the approach in [12, 19]. The power consumption by router model type is given in Table 7, while the power consumption breakdown for the different router components is given in Table 4.

From Table 4, the forwarding engine and the control plane together account for 44.5 % of the power consumption in a router. This is the power consumed to run the routing protocol and related processing needs in a router. Memory access consumes 7.0 % of power used by the router.

In this study, power usage (PU) cost will be determined for the different types of router. And in each type of router, the power usage for memory and

Table 7. Power consumption of router types. Source: [12, 14]

Types	Models	Power consumption/hour
Access Router (AR)	Cisco 7603	1.0 KW
Distribution Router (DR)	Cisco 12816	6.0 KW
Backbone Router (BB)	Cisco CSR-1	10.0 KW

routing operations will be determined. From Table 7 the power consumed by a BB router is 10.0 KWatts, by a DR router is 6.0 KW and by an AR router is 1.0 KW, assuming that the routers are running OSPF.

Using Eq. (3) the complexity ratio between OSPF and TRP routers in the AT&T ISP network was computed for every router. The plot for this is given in Fig. 8 for the 12,000 routers for which this was computed. In the x axis is the incremental number of routers for which complexity ratio was computed.

Memory Usage by Routing Tables. The memory or storage needs in router depends on the routing table sizes and also on the length of the network address that is being stored. For a router x running OSPF, in a domain that has L links, if x is a BB router then $L(x)$ is all the links in the AT&T network. If x is a DR or AR router then $L(x)$ is the number of links in the POP of router x. Assuming that we are storing IPv4 addresses, to store each address we need 32 bits.

With networks running TRP the routing table size is given by N, which is the number of direct neighbors of a router x. The TRA size will have a maximum value of 12 bits [26]. Hence the ratio of memory storage required to store routing tables populated by TRP to OSPF will be given by

$$RT_{ratio}(x) = \frac{N(x) \times 12bits}{L(x) \times 32bits} \tag{8}$$

where $L(x)$ is *all links in AT&T* when x is a BB router and $L(x)$ is *all links in a POP X*, where x is DR and AR.

$RT_{ratio(x)}$ was computed for 12,000 routers individually and plotted in Fig. 7.

Power Usage (TRP Routers). PU at current OSPF routers is given by

$$PU_{OSPF} = (BB_{OSPF} \times 10.00) + (DR_{OSPF} \times 6.00) + (AR_{OSPF} \times 1.00) \tag{9}$$

From Tables 4 and 7, at a BB router running OSPF, PU_{mem}, the power consumed for memory storage and PU_{rt}, the power consumed for routing operation can be calculated as given in Eq. (10).

$$PU_{mem} = 0.07 \times 10.0$$
$$PU_{rt} = 0.445 \times 10.0 \tag{10}$$

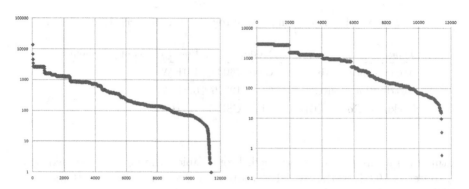

Fig. 7. Routing table ratio between OSPF and TRP in the AT&T network

Fig. 8. Complexity ratio between OSPF and TRP in the AT&T network

Hence power consumed by rest of the router components is given by

$$PU_{else} = 0.485 \times 10.0 \tag{11}$$

Similarly at a DR router running OSPF,

$$PU_{mem} = 0.07 \times 6.0$$
$$PU_{rt} = 0.445 \times 6.0$$
$$PU_{else} = 0.485 \times 6.0 \tag{12}$$

And for an AR router running OSPF

$$PU_{mem} = 0.07 \times 1.0$$
$$PU_{rt} = 0.445 \times 1.0$$
$$PU_{else} = 0.485 \times 1.0 \tag{13}$$

To compute the power consumption in a TRP router, the PU_{else} is maintained to be the same as an OSPF router. Power consumed to support routing operation and to store routing tables will be computed using C_{ratio} and RT_{ratio} respectively. Hence power usage at a BB router running TRP is given by

$$PU_{BB\_trp}(bb) = PU_{else} + PU_{rt} \times C_{ratio}(bb)$$
$$+ PU_{mem} \times RT_{ratio}(bb) \tag{14}$$

where bb is any BB router under the new TRP deployment. C_{ratio} and RT_{ratio} change with the type of router namely BB, DR and AR. Similarly,

$$PU_{DR\_trp}(dr) = PU_{else} + PU_{rt} \times C_{ratio}(dr)$$
$$+ PU_{mem} \times RT_{ratio}(dr) \tag{15}$$

and,

$$PU_{AR\_trp}(ar) = PU_{else} + PU_{rt} \times C_{ratio}(ar)$$
$$+ PU_{mem} \times RT_{ratio}(ar) \tag{16}$$

· From Eq. (9) and summing up for all the BB, DR and AR routers running TRP, the power consumption and power usage cost was computed on a per year basis and is given in Table 8. The cost per hour used in the computations was $60/MWh [13, 29].

Table 8. Total yearly estimated power usage cost and reduction

	Power consumption	PU cost
BB	34,076 MWh	$2,045K
DR	336,121 MWh	$20,167K
AR	40,462 MWh	$2,428K
Total (before)	410,659 MWh	$24,649K
Cost ratio	14.02 %	
Total (after)	57,574 MWh	$3,454K

Table 8 also shows the estimated reduction in yearly cost of PU($60/MWh) before and after the transition. The BB routers currently consume 34,076 MWh, hence the PU cost by BB routers in the AT&T network is $2,045K. The PU cost by all the routers in the AT&T network in the US can thus be calculated to be $24,649K. If TRP routers were used, this cost would be cut down to 14.02 % of the current cost. The PU cost would be $3,454K resulting in a total savings of $21,195K per a year.

Figure 9 shows the PU cost ratio of each POP after the transition. The PU cost ratio reflects the reduction in the PU cost. For example, PU cost of Chicago, IL POP is 17.15 % of the original PU cost before the transition. If we summed up for all POPs, the reduction in PU cost for the AT&T network would be 14.02 % of the current PU cost.

5   The Transition Study

For deployment purposes of a new technology it is also important to study the energy saving and costs during a typical transition scenario. Replacements and substitution can happen over several weeks and costs incurred during this period and the process is important to any business organization that is deploying the technology before the actual deployment/substitution takes place. Hence in this study we calculate the power usage cost during a transition scenario. The transition was conducted POP by POP assuming that each POP operates independently.

The routers were replaced starting from the BB routers, and then to the DR routers and lastly the AR routers in a POP. For this case study, it was decided to replace a maximum of 50 routers in a POP per week. If there are more than 50 DRs or ARs in a POP, the replacement order will depend on the number

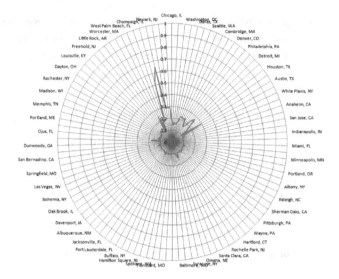

Fig. 9. Cost ratio of PU before and after transition(POP)

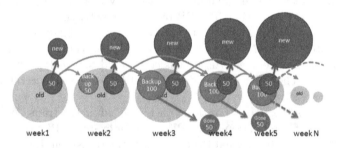

Fig. 10. Transition scenario in a POP

of hops that a router is from a BB router, *i.e.* routers closer to the BB router will be replaced first. Since TRA allocation starts from the top tier with TRP, the transition scenario also followed a top-to-bottom replacement approach in the network topology. Therefore, router replacement starts from BB routers in each POP. After replacement of old routers, they were still kept as backup for a period of at least 2 weeks for the sake of risk management.

Figure 10 shows the week by week transition where 50 of the old routers are replaced in a POP in a week. The number of new routers increase over the weeks and the number of old routers decrease. As it was assumed that each POP works independently, all POPs in the AT&T network can transition simultaneously.

Operating costs during the transition was calculated on a weekly basis. After all BB routers are replaced in a POP, DR routers will be replaced. The replacement order of DR and AR routers in POP was based on the topology in each POP. The shortest path to BB router was identified for each of the DR and AR routers, and routers that had shorter hops to BB were replaced first.

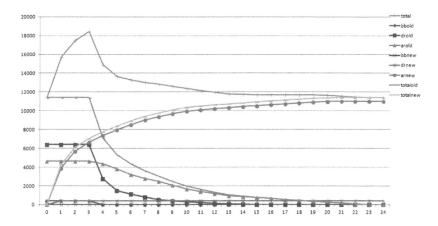

Fig. 11. Number of each types of routers during transition

The numbers and locations of BB, DR, and AR routers in a POP in the AT&T network was already determined and was used in the transition study.

5.1 Router Substitution Process During Transition

Figure 11 shows result of router types as they are transitioned over weeks. All old routers are replaced by new routers by the 24th week. The numbers of old routers do not change in the first 3 weeks because old routers will be kept for a 2-week period after being replaced by the new routers. Thus, total number of routers increase in the first 3 weeks. After the first 3 weeks, the number of old routers starts to decrease. Due to the router substitution in terms of their capacity most old routers were eventually replaced by ARs.

5.2 Power Usage Cost During Transition

Power Usage (PU) cost during transition is shown in Fig. 12. Weekly PU cost before the transition (week 0) is \$473K and PU cost after the transition (week 24) is \$66K, and the peak PU cost is \$517K in week 3. As most of the old BB and DR routers are replaced by new AR routers in the first few weeks, the PU cost reduces significantly by week 5.

6 Conclusion

A modular routing architecture and routing protocol based on this architecture was developed and evaluated for its performance efficiency in terms of reduced routing table sizes and routing operation complexity. In this article the energy efficiency of adopting this routing protocol has been evaluated in comparison with the current routing protocol used within ISP networks namely the Open

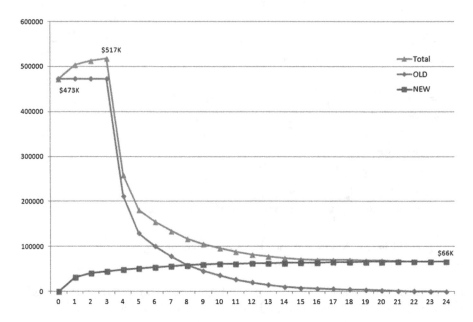

Fig. 12. PU cost ($) during transition

Shortest Path First (OSPF) protocol. The new modular routing protocol named the tiered routing protocol offers significant reduction in the routing operational and memory storage needs in a router. This huge reduction reflects in significant reduction in energy used by routers implementing the new protocol. To the best of our knowledge, this study is the first of its kind where statistics based on a real ISP network namely the AT&T ISP network in the US was used to investigate the reduction in operational complexity and memory storage needs for the different types of routers and this was evaluated for all the POPs and routers in this ISP network. The comparative power usage studies show the huge potential of the proposed solution to reduce energy concerns. The studies used realistic market prices and power usage by typical routers used in ISP networks to calculate several comparative indices that allowed calculation of the power usage in a potential new router. Thus the work provides a framework for such future investigative studies.

References

1. Arkansas Cisco Price List. https://www-304.ibm.com/easyaccess3/fileserve?contentid=107260. Accessed 30 Jan 2016
2. Emulab Network Emulation Testbed. http://www.emulab.net
3. Internet Research Task Force Routing Research Group. http://trac.tools.ietf.org/group/irtf/trac/wiki/RoutingResearchGroup. Accessed 30 Jan 2016
4. National Science Foundation Future Internet Architecture Project. http://www.nets-fia.net/. Accessed 30 Jan 2016

5. NSF grants,: #0739362, #1036636, #0832008
6. NSF NeTS FIND Initiative. http://www.nets-find.net. Accessed 30 Jan 2016
7. Quagga Software Routing Suit. http://www.nongnu.org/quagga/. Accessed 30 Jan 2016
8. The Network of the Future Projects of EU FP7. http://cordis.europa.eu/fp7/ict/future-networks/. Accessed 30 Jan 2016
9. Alderson, D., Li, L., Willinger, W., Doyle, J.C.: Understanding Internet topology: principles, models, and validation. IEEE/ACM Trans. Networking **13**(6), 1205–1218 (2005)
10. Allman, M., Christensen, K., Nordman, B., Paxson, V.: Enabling an energy-efficient future internet through selectively connected end systems. In: HotNets (2007)
11. Argyraki, K., Baset, S., Chun, B.G., Fall, K., Iannaccone, G., Knies, A., Kohler, E., Manesh, M., Nedevschi, S., Ratnasamy, S.: Can software routers scale? In: Proceedings of the ACM Workshop on Programmable Routers for Extensible Services of Tomorrow, pp. 21–26. ACM (2008)
12. Baliga, J., Ayre, R., Hinton, K., Tucker, R.: Photonic switching and the energy bottleneck. In: Photonics in Switching, vol. 2007, pp. 125–126 (2007)
13. Bolla, R., Bruschi, R., Carrega, A., Davoli, F., Suino, D., Vassilakis, C., Zafeiropoulos, A.: Cutting the energy bills of Internet service providers and telecoms through power management: an impact analysis. Comput. Netw. **56**(10), 2320–2342 (2012)
14. Bolla, R., Bruschi, R., Davoli, F., Cucchietti, F.: Energy efficiency in the future Internet: a survey of existing approaches and trends in energy-aware fixed network infrastructures. IEEE Commun. Surv. Tutor. **13**(2), 223–244 (2011)
15. Bolla, R., Bruschi, R., Ranieri, A.: Green support for PC-based software router: performance evaluation and modeling. In: IEEE International Conference on Communications, ICC 2009, pp. 1–6. IEEE (2009)
16. Ceuppens, L., Sardella, A., Kharitonov, D.: Power saving strategies and technologies in network equipment opportunities and challenges, risk and rewards. In: International Symposium on Applications and the Internet SAINT 2008, pp. 381–384. IEEE (2008)
17. Chiaraviglio, L., Mellia, M., Neri, F.: Minimizing ISP network energy cost: formulation and solutions. IEEE/ACM Trans. Networking (TON) **20**(2), 463–476 (2012)
18. Cianfrani, A., Eramo, V., Listanti, M., Polverini, M.: An OSPF enhancement for energy saving in IP networks. In: 2011 IEEE Conference on Computer Communications Workshops (INFOCOM WKSHPS), pp. 325–330. IEEE (2011)
19. Gianoli, L.G.: Models and algorithms for energy saving in IP networks (2010)
20. Gombiner, J.: Carbon footprinting the Internet. Cons.: J. Sustain. Dev. **5**(1), 119–124 (2011)
21. Gupta, M., Singh, S.: Greening of the Internet. In: Proceedings of the 2003 Conference on Applications, Technologies, Architectures, and Protocols for Computer Communications, pp. 19–26. ACM (2003)
22. Moy, J.: RFC 1245 OSPF Protocol Analysis. IETF Internet Standard, RFC 1245 (1991)
23. Nedevschi, S., Popa, L., Iannaccone, G., Ratnasamy, S., Wetherall, D.: Reducing network energy consumption via sleeping and rate-adaptation. In: NSDI, vol. 8, pp. 323–336 (2008)
24. Nozaki, Y., Bakshi, P., Shenoy, N.: A novel approach to interior gateway routing. Int. J. Adv. Netw. Serv. **6**(3–4), 208–219 (2013)
25. Nozaki, Y., Bakshi, P., Tuncer, H., Shenoy, N.: Evaluation of tiered routing protocol in floating cloud tiered Internet architecture. Comput. Netw. **63**, 33–47 (2014)

26. Nozaki, Y., Tuncer, H., Shenoy, N.: A tiered addressing scheme based on a floating cloud internetworking model. In: Yu, H., Vaidya, N.H., Srinivasan, V., Choudhury, R.R., Aguilera, M.K. (eds.) ICDCN 2011. LNCS, vol. 6522, pp. 382–393. Springer, Heidelberg (2011)

27. Panigrahy, R., Sharma, S.: Reducing TCAM power consumption and increasing throughput. In: 10th Symposium on High Performance Interconnects, Proceedings, pp. 107–112. IEEE (2002)

28. Penttinen, A.: Green networking-a literature survey. Technical report. Department of Communications and Networking, Aalto University (2011)

29. Qureshi, A., Weber, R., Balakrishnan, H., Guttag, J., Maggs, B.: Cutting the electric bill for Internet-scale systems. ACM SIGCOMM Comput. Commun. Rev. **39**(4), 123–134 (2009)

30. Ravikumar, V., Mahapatra, R.N.: TCAM architecture for IP lookup using prefix properties. IEEE Micro **24**(2), 60–69 (2004)

31. Roberts, L.G.: A radical new router. IEEE Spectr. **46**(7), 34–39 (2009)

32. Spring, N., Mahajan, R., Wetherall, D., Anderson, T.: Measuring isp topologies with rocketfuel. IEEE/ACM Trans. Networking, **12**(1), 2–16 (2004)

33. Yoo, S.: Energy efficiency in the future Internet: the role of optical packet switching and optical-label switching. IEEE J. Sel. Top. Quantum Electron. **17**(2), 406–418 (2011)

34. Yu, H.: A memory-and time-efficient on-chip TCAM minimizer for IP lookup. In: Design, Automation and Test in Europe Conference and Exhibition (DATE), pp. 926–931. IEEE (2010)

Efficient Security Policy Reconciliation in Tactical Service Oriented Architectures

Vasileios Gkioulos[1(✉)] and Stephen D. Wolthusen[2,3]

[1] Norwegian Information Security Laboratory,
Norwegian University of Science and Technology, Gjøvik, Norway
vasileios.gkioulos@ntnu.no
[2] Norwegian Information Security Laboratory,
Gjøvik University College, Gjøvik, Norway
[3] School of Mathematics and Information Security, Royal Holloway,
University of London, Egham, UK
stephen.wolthusen@ntnu.no

Abstract. Tactical mobile ad-hoc networks are likely to suffer from highly restricted link capacity and intermittent connectivity loss, but must provide secure access to services. The conditions under which services may be accessed and which security requirements must be maintained will vary dynamically, and local policies will hence change on a per-node basis even when starting from a common baseline such as when nodes obtain new information.

In this paper we describe a mechanism allowing structured security policies to incorporate such local changes but to efficiently reconcile across tactical SOA networks, allowing the derivation of policy decisions as precomputed Horn clauses or directly reasoning over a description logic fragment. This mechanism minimises the communication overhead compared to earlier work whilst maintaining policy integrity, thereby allowing security policies to adapt to resource and network constraints and other local knowledge such as node compromises and blacklisting.

Keywords: Ad hoc network · Reconciliation · Security · Security policies · Tactical network

1 Introduction

Tactical networks are mobile wireless ad-hoc or mesh networks with frequently severely limited resources and also subject to loss of connectivity owing to aspects ranging from mobility to adversaries jamming. The use of service-oriented architectures (SOA) allows nodes to invoke and dynamically configure services depending on factors including service availability. However, whilst nodes in such a network may commence a mission with a consistent security policy and knowledge of the respective local state and environment, this will evolve over time. A security policy here relies upon a knowledge base for the target domains, node capabilities and constraints, allowing the dynamic inclusion

© Springer International Publishing AG 2016
R. Doss et al. (Eds.): FNSS 2016, CCIS 670, pp. 47–61, 2016.
DOI: 10.1007/978-3-319-48021-3_4

of local state and environmental knowledge for the on-line selection and configuration of security controls. In SOA, this allows the dynamic invocation and orchestration of services, selecting the node for which services or service choreography may be optimal, and which security controls and mechanisms such as protocols and algorithms are required or supported.

Earlier studies [1,2] investigated tactical SOA, defining suitable protection goals, security requirements and policy design preconditions in consistence to the identified constraints. Such constraints include the required scalability and dynamic adaptation of the security mechanisms, in addition to the inherently requisite support of heterogeneity, functional diversity and cooperativity across the tactical nodes. Ontologies have been identified as a suitable mediator towards the realisation of security requirements in distinct domains. [3–8]. Extending this paradigm to tactical SOA [9], the aforementioned preconditions have been translated into security structural and functional requirements. These, necessitated the realization of robust yet flexible protection mechanisms, able to dynamically adapt to the environmental alterations, maintaining support over the defined set of security goals. Thus, the same study suggested a security policy framework dedicated to tactical SOA based on Web Ontology Language (OWL), as OWL offers the required scalability and distributed operation, providing sufficient expressive power for capturing and reasoning over the underling semantics [10–14].

Yet, the functional limitations of tactical nodes render the mere replication of security policies infeasible, while the implemented security mechanisms cannot rely on centralised configurations, since continuous connectivity towards a security dedicated entity cannot be reassured. Thus, a mechanism for the efficient distribution of ontologically defined security policies over tactical SOA has been developed earlier [15]. As specified previously, the distributed security policies must be able to adjust and respond to the continuous alterations of the tactical environment, transitioning between consistent states. This necessitates the incorporation of dynamic semantics within the security policy, which can cause local divergences regarding its scope or context. Such divergences can lead to policy inconsistency and node antagonism, affecting network performance in various terms, including service delivery. Thus, the reconciliation of occurring discrepancies among the distributed ontologies is required.

This paper presents our findings in respect to the reconciliation of singular (or a priory mapped in the case of coalition environments) distributed ontological security policies for tactical SOA, focusing on the mission execution stage. The objective of this study is to achieve this, while minimizing the security induced overhead, both in terms of computational complexity and bandwidth consumption. Section 2 presents related work, while at Sect. 3 the main concepts of the previously developed tactical policy model are briefly presented. Section 4 includes our findings regarding the nature of the occurring divergences, aiming to minimize the complexity of the reconciliation mechanism and the size of transmitted messages. Consequently, Sect. 5 presents the functionality of the components constituting such a mechanism (in respect to the content of Sect. 4), while Sect. 6 includes the formalisation of these ordered functional elements in algorithmic form.

2   Related Work

The resolution of heterogeneity by semantic alignment of distinct ontologies corresponds to ontology mapping, which is a mature area of research both for static and dynamic ontologies [16–27]. Yet, these methodologies are not suitable for the specifics of tactical SOA, since their definition was targeted on dissimilar domains and the corresponding constraints were not addressed. Some of these mechanisms aim to the mapping of distinct ontologies, focusing on creating a linking dictionary which is not necessary in the case of national or coalition operations, regarding the mission execution stage. Additionally, no communication restrictions are considered, requiring multiple transactions or the transmission of the entire ontology. Furthermore, some mechanisms allow residue unresolved divergences, or require the initiation of a complete mapping cycle, either periodically or at any time that a differentiation is detected.

Focusing on military applications, Bakillah et al. [28] provided a flexible semantic mediation mechanism for heterogeneous sensor data. Yet, despite the nature of sensor networks, no communication restrictions have been considered. Furthermore, Besana and Robertson [29] suggested the incorporation of service choreography statistics, for the minimization of the ontology mapping problem, over open and distributed environments. Yet, the utilization of such mechanisms for security dedicated ontologies, may allow residue or pending divergences that, thought not relevant for a given transaction, can remain unresolved and be subject to adversarial exploitation. Moreover, Trivellato et al. [8] presented a mapping mechanism at the security domain of maritime coalition environments. That study focus on the mapping of not singular but distinct security policies, and due to the nature of maritime nodes dissimilar communication constraints are considered. Finally, Muthaiyah and Kerschberg [30] also focus on the security domain of ontology mapping, but the proposed mechanism does not allow operation over distributed and constrained environments, since it requires the exchange of the entire ontology, for every mapping cycle.

Khattak et al. [31–34] proposed an ontology mapping mechanism where only the altered semantics are exchanged and reconciled by a centralised mapping system. Such an approach has been proven to provide increased reconciliation efficiency, in terms of time and computational power consumption. Thus, adopting this paradigm across tactical SOA, we seek to satisfy the discrete operational (e.g. distrubuted operation), security (e.g. increased reconciliation confidence) and functional (e.g. bandwidth consumption) requirements, as presented below.

3   Security Policy Formulation and Reasoning

In this section the architecture, formal representation and distribution mechanisms for the examined security policies are briefly presented, according to the results of our previous studies [1,2,9,15]. This is crucial for the identification of the components and functionalities, required for the investigated reconciliation mechanism.

3.1   Security Policy Architecture

Supporting the requisite functionalities and dynamic service orchestration over tactical SOA, necessitates the fine-grained conceptualization of the constituent network elements, in correspondence to the anticipated processes and operational requirements. Exploiting the expressive power of OWL, such a mechanism can be defined as presented, in small scale, at Fig. 1.

The anticipated processes and requirements are conceptualised by the unambiguous representation of the tactical domains (Such as planning, management, detection, intelligence and diligence) and operational capabilities (Such as communication, core, inter-domain and application). The intersection of these two elements corresponds to a predefined set of required actions, which can be visualised in a three dimensional space, with a non uniform distribution. Concurrently, each action is governed by a dynamically selected set of prioritized rules. These are constructed with increasing complexity and preciseness, supporting both the cooperative and standalone functionality of the tactical nodes, in conjunction with their functional characteristics and available resources. These rules also incorporate and serve as links towards the aforementioned static and dynamic properties of the constituent network elements (Namely services, information, network, radios, nodes and subjects).
Thus:

$$Individual\_Domain \cap Individual\_Capability = \{Individual\_Action\_A(k),$$
$$Individual\_Action\_A(k+1), ..., Individual\_Action\_A(k+j)\} \quad (1)$$

where:

$$Individual\_Action\_A(k) \hat{=} \{Rule\_A[k(z)], Rule\_A[k(z+1)], ..., Rule\_A[k(z+i)]\} \quad (2)$$

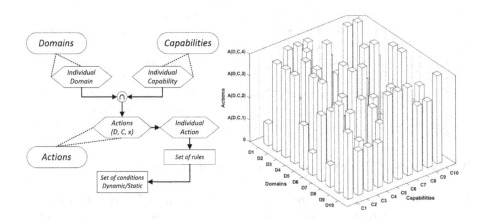

Fig. 1. Outline of security policy structure.

3.2   Formal Representation

The formal representation of the aforementioned elements is achieved with the utilization of the constructors, provided by the selected description logic (DL) fragment for the formulation of appropriate unary and binary predicates. In order to achieve precise capturing of the required concepts, the selected DL-fragment must be based on *ALC*, but also support role hierarchies and inclusion, inversion, nominals, functionality properties and qualified cardinality restrictions. *SHOIN(D)* has been identified as a suitable DL-fragment, but more lightweight fragments can also be utilised for optimization purposes. The tactical terminology is constructed within the corresponding T-box, in terms of acyclic and unique concept definitions, as a set of sufficient and necessary conditions. Consequently, constructing expressions similar to those presented at Eqs. 3, 4 and 5, allows to exploit the expressive power of DL in order to gradually structure all the individual concepts, across the distinct tactical domains.

$$Terminal \equiv individual \sqcap \exists has\_Terminal\_ID. \perp \qquad (3)$$

$$Local\_Provider \equiv Terminal \sqcap \exists Has\_Operational\_Group.OG2$$
$$\sqcap \exists Has\_Status.Online \sqcap \exists Has\_Functionality.SP \qquad (4)$$

$$Available\_Service \equiv Service \sqcap \leq 1 Has\_Local\_Provider \qquad (5)$$

Additionally, A-box is oriented to instance identification, where concept and role assertions are utilized in order to specify a given individual as an instance of a specific concept, as presented at Eq. 6.

$$Concept \ assertion$$
$$File \sqcap Video(Message\_x) : Message\_x \ is \ a \ video \ file$$
$$Role \ assertion$$
$$hasSource(Message\_x, Terminal\_y) : Terminal\_y \ is \ the \ source \ of \ Message\_x$$
$$(6)$$

3.3   Partitioning and Distribution

Assuming a tactical security policy, the efficient distribution of the corresponding ontological structure across the deployed actors, requires the evaluation of various network and operational parameters, as presented in Fig. 2. Incorporating these elements into the distribution mechanism at the mission preparation stage, allows the responsibility allocation for the required actions of a given tactical operation, taking under consideration the structure of the policy, the characteristics and expected behaviour of the tactical nodes, alongside the required dynamic functionalities. This allows the minimization of the policy responsibility overlap across the deployed actors, maintaining their capacity for standalone operation.

Thus, evaluating the syntactic and structural complexity of the ontological structure, in combination with the incorporated dynamic attributes, allows its

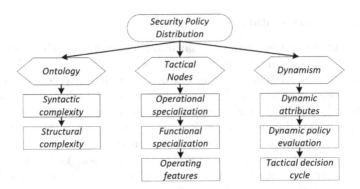

Fig. 2. Governing parameters of security policy distribution, over tactical SOA.

partitioning and distribution across various node groups, organised based on their required operational/functional behaviour and their operating features. Additionally, the incorporation of dynamic policy evaluation mechanisms and a tactical decision cycle, allows the extended partitioning of policy decisions, when they are utilized during mission execution. The security policy distribution is a prerequisite of tactical SOA, but as presented earlier it raises the question of reconciling the occupying divergences.

4   The Characteristics of Occurring Divergences

Investigating the nature of occurring divergences, four common types of tactical operations have been analysed and simulated, namely (i) Tactical convoy, (ii) Reconnaissance Surveillance and Target Acquisition (RSTA), (iii) Intervention patrol and (iv) Medical Evacuation (MEDEVAC). Each operation was partitioned into a variety of use cases (e.g. Blue force tracking and common operational picture distribution, injection of high mobility nodes, improvised explosive device (IED) detection and report, interoperability with police forces) including detailed episodes (e.g. Addressed request/reply, multi-hop service invocation, service discovery and node isolation). This analysis was based on a security policy (see Sect. 3) constructed using the DL fragment *ALCHIF(D)* as depicted at Fig. 3, while this core ontological model was adjusted to the specifics of each tactical operation.

Assuming the simplified scenario presented at Table 1, a divergence at the local knowledge of two nodes (Node_A, Node_B) regarding the status of a sensor attached to a vehicle is presented. In this scenario, during the mission execution stage, hostile forces achieve local prevalence at a given Area of Operation (AoO4), thus the trust level of the locally deployed sensors is automatically degraded. Sensor_09134 is responsible for gathering local blue force tracking data (At the tactical team level), incorporating them into a low resolution local aerial photo and transmitting the output periodically across the network. Node_A becomes aware of the final position of Sensor_09134 (AoO4) thus limits the

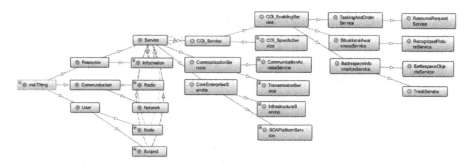

Fig. 3. Investigated security policy/ontology (Parts of the service sub-tree are expanded.)

Table 1. Simplified differentiation scenario from the intervention patrol simulation set.

Local knowledge at Node_A	Local knowledge at Node_B
Information (Message)	Information (Message)
has_Classification(MSG_x, Top_Secret)	has_Classification(MSG_x, Top_Secret)
has_Nature(MSG_x, Blue_Force_Tracking)	has_Nature(MSG_x, Blue_Force_Tracking)
has_Type(MSG_x, Image_jpg)	has_Type(MSG_x, Image_jpg)
has_Size(MSG_x, 200)	has_Size(MSG_x, 200)
has_Source(MSG_x, Sensor_09134)	has_Source(MSG_x, Sensor_09134)
⋮	⋮
Node (Sensor_09134)	Node (Sensor_09134)
has_State(Sensor_09134, Active)	has_State(Sensor_09134, Active)
has_Trust(Sensor_09134, 20)	has_Trust(Sensor_09134, 87)
has_Location(Sensor_09134, AoO4)	has_Location(Sensor_09134, AoO7)
⋮	⋮

trust level according to the security policy, while Node_B maintains the previous update (AoO7) incorporating no alterations and treating information based on the previous status. Similar scenarios can occur in the case of a communication disruption, when a group of nodes (Node_A, Sensor_09134) reconnects with other parts of the network and continues to operate, exchanging information. This scenario refers to a simplified divergence, yet the extent, content and impact of such alterations can vary according to the context of the tactical operation and the structure of the security policy.

The results of our analysis can be summarised as:

– Strict syntactic, terminological and semiotic homogeneity is maintained [35], since the distributed local ontologies are consistent with respect to the central ontological model. Thus, no policy management in respect to conflict resolution (e.g. vocabulary translation, rule deconflictation) between distinct policies is required by the reconciliation mechanism.

– Divergences occur only due to conceptual heterogeneity. This is further restricted since the local ontologies operate within only two dimensions of context dependent representation, namely partiality and perspective, but not in respect to approximation [36]. Hence, minimising granularity negotiations and the corresponding message transmissions.

– Approximation differences are utilized across the defined governing rules of each available action. Thus, it is locally maintained in order to provide dynamic policy adaptation to the tactical network dynamics, without increasing the security induced overhead during multi-party policy reconciliation.

– The elements of the ontological structure that can be affected by such alterations, occur only within the values of defined object and data properties, while their respective ranges and domains remain unaffected. Additional centralized revisions that may require alterations within classes, individuals and SWRL rules, would require a global policy update, which will have to incorporate alternative and more costly mechanisms.

– The type of allowed alterations includes only the modification (Revision) of the identified elements, since their extension (Addition) or reduction (Deletion) would correspond to the privilege allocation to each individual node, of modifying the tactical security policy. Thus, only the revision of the identified elements should be expected and allowed, supporting the adaptation of the extracted policy decisions, based on the evolution of the dynamic semantics across the network.

These findings allow the simplification of the developed reconciliation mechanisms and the minimization of the network resources allocated for this purpose. No terminology or structural negotiations are required, while the divergence targets and types can be considered static. The impact, as visualised at Fig. 4, is located both to the size and complexity of Δ (Divergence to be transmitted and reconciled), significantly minimizing the consumption of bandwidth (When treated as transmitted datum) and computational power (When treated as data at rest).

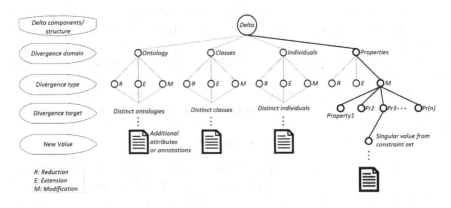

Fig. 4. Ontology divergence mapping tree.

Centralised ontology mapping methodologies require the transmission of the entire ontology either bilaterally to a reconciliation dedicated entity, or unilaterally among the dissident nodes. Change reconciliation based mechanisms provide increased efficiency in terms of elapsed time and computational complexity by referring only to the altered elements. This is achieved by the construction and algorithmic support of a complete *Ontology divergence mapping tree* (ODMT), as presented in Fig. 4, and the communication of the altered elements in XML or encoded format. The presented findings allow further improvement by permitting the reduction of ODMT to the property branch, providing an absolute minimum of Delta components (Divergence target + New Value). The satisfaction of network and security requirements necessitates the incorporation of additional attributes and annotations (e.g. Divergence source and previous hops ⇒ auditing and non repudiation, Time stamp and precedence ⇒ freshness and prioritization, Divergence trust ⇒ reconciliation confidence). Yet, the initial reduction of ODMT has a significant impact in terms of bandwidth and computational efficiency.

5 Identification of Required Elements and Functionalities

The reconciliation of ontologically defined security policies is closely related to ontology mapping mechanisms. Yet, as presented earlier, such solutions are constructed for operation within domains with distinct requirements and constraints to tactical SOA. Zablith et al. [17] described an ontology evolution cycle comprising of five main steps, presenting the corresponding existing mechanisms (Detecting the need for evolution ⇒ Suggesting changes ⇒ Validating changes ⇒ Assessing impact ⇒ Managing changes). The developed security policy reconciliation mechanism required the addition of a communication step, responsible for the adaptation to the characteristics of tactical SOA.

Through the analysis and simulation of the aforementioned tactical operations, the requisite functionalities of the *Communication* step have been identified. Thus, such a mechanism is required to minimize:

- The knowledge propagation time
- The size of transmitted elements
- The number of involved nodes
- The complexity of the transmitted elements
- The number of interactions

Additionally, auditing, prioritization and roll back capabilities must be enabled, maintaining increased reconciliation confidence. The presented results regarding the nature of occurring divergences, can be efficiently utilized in order to minimize the size and complexity of transmitted elements, while the additional requirements are attained by the use of appropriately constructed mechanisms.

OWL operates over the open world assumption, which is required by the functional characteristics of the defined security policy. Yet, it is possible to enforce closed world assumption during the construction of the policy by the definition of explicit constraints. Thus, data driven evolution is possible. This can be achieved by recording the Δ caused by the various data sources (services, terminals, users) and initiate the reconciliation either as event driven (a session related policy reconciliation) or when the Quality of Service (QoS) mechanisms signal that the required resources have become available. The developed mechanisms for the achievement of the aforementioned requirements are:

1. Local ontology (Fragment of the global ontology/policy (Sect. 3))
2. Local node assignment list (Fragment of a global node assignment list, responsible for the identification of the subset of nodes, which incorporate the altered element.)
3. Local change ontology (Maintains a copy of locally sensed and enforced changes for audit and roll back purposes).
4. Criticality/ timeliness measure (For prioritization purposes)
5. Archive of requested changes (Maintains a copy of externally requested changes for audit and roll back purposes).
6. Δ (It includes the altered element, and various characteristics of the alteration, such as justification, time, actor.)

The global node assignment list (G-NAL) operates as a responsibility database, used for the initial partitioning and distribution of the security policy across the deployed actors. The *local node assignment list* is a fragment of G-NAL that during a policy reconciliation at the mission execution stage is used in order to minimise the number of involved nodes and interactions, to the minimum acceptable subset of recipients/transmitters that provide sufficient reconciliation confidence. This mechanism has been constructed with minimum complexity using the *SF(D)* DL fragment, as presented in Fig. 5 with the use of a transitive object property (e.g. Makes_Use_Of) between the deployed assets (Nodes), the required actions and the existing object or data properties, providing a mapping between the nodes and the possibly altered attributes. Querying this ontology (e.g. Uses value ServiceStatus and HasOG some string) provides a list of nodes that have to be updated once a change is detected locally (e.g. on the ServiceStatus), or the list of nodes that belong to the same Operational group and are expected to transmit update requests (Used for Get_Recipients, Get_Requesters, Get_Properties of Sect. 6).

Additionally, the required criticality and timeliness measures have been attached to the properties of this mechanism, in the form of data properties, in order to provide prioritization of the reconciliation requests in a congested environment. Criticality measures follow military precedence designators as flash (e.g. Intrusion detection, compromised node), Immediate (e.g. Trust level update), Priority (e.g. Local service provider, service registry update) and routine (e.g. Location update, service status update). Prior to transmission the extracted set of Δ are initially classified in respect to their criticality (Maximum first), and consequently based on their life cycle (Minimum first), while the corresponding update requests are transmitted in accordance to the available network resources (Used for Sort_Changes of Sect. 6).

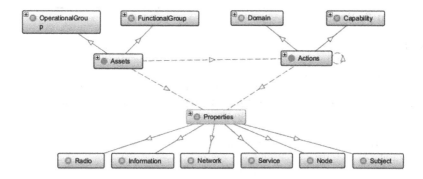

Fig. 5. Structure of node assignment list.

Khattak et al. [31–34] presented a flexible and robust mechanism for the construction of the required local change ontology (which is also suitable for the inbound oriented, archive of requested changes), combined with the capability of extracting the required Δ and its use for the mapping of dynamic ontologies. These mechanisms can be modified in order to serve only the identified components, based on the nature of occurring divergences (Sect. 4). Thus, being adequately lightweight in order to serve under the tactical constraints and update the locally stored ontologies given a specific Δ (Used for Update_Local_Policy of Sect. 6).

6 Policy Reconciliation Mechanism

A reconciliation mechanism has been developed based on the presented elements, in order to provide the aforementioned functionalities. The corresponding algorithms for the formalization of the ordered functionality sets are presented bellow.

Algorithm 1. Transmitter

1: ▷ Identify that a set of properties evolved locally. Related "Change_Detection" alert coming from a Meta-data Handler or Contextual Monitoring local service.
2: **if** Change_Detection == TRUE **then**
3:    ▷ Incorporate changes into the local security ontology. (Existing mapping mechanisms).
4:    **Update_Local_Policy**(Δ,Security_Ontology)
5:    ▷ Incorporate changes into the local change ontology (Existing mapping mechanisms).
6:    **Update_Local_Policy**(Δ,Local_Change_Ontology)
7:    ▷ Query the local node assignment list for the corresponding list of nodes (As described).
8:    (Recipients_List)**Get_Recipients**(Δ)
9:    ▷ Apply criticality and prioritization measures (As described).
10:    (Δ')**Sort_Changes**(Δ)
11:    ▷ Send to QoS Mechanisms for transmission to the list of recipients.
12:    **Send_QoS**(Δ', Recipients_List)
13: **end if**

Algorithm 2. Receiver

1: ▷ Receive the first update request for a specific property.
2: **if** Receive_Update_Request.$\Delta'[x]$ == TRUE **then**
3:     ▷ Store changes to archive of requested changes (Existing mapping mechanisms).
4:     **Update_Local_Policy**(($\Delta'[x]$), AoRC)
5:     ▷ Query local node assignment list for expected requests. (Similarly to Get_Recipients, it provides a list of nodes that belong to the same operational group, thus sensed the property alteration and are expected to transmit similar requests).
6:     (Requester_List)**Get_Requesters**($\Delta'[x]$)
7:     ▷ Query local node assignment list for lifetime and criticality measures of $\Delta'[x]$ (As described.)
8:     (T$\Delta'[x]$, C$\Delta'[x]$)**Get_Properties**($\Delta'[x]$)
9:     ▷ Request estimated list of requests from QoS mechanisms. (Identify requesters who have the resources to transmit requests within the T$\Delta'[x]$)
10:     (*Requester_List'*)**QoS_Estimation**(Requester_List, T$\Delta'[x]$)
11:     ▷ Wait for the expected update requests.
12:     **while** T$\Delta'[x]$!=0 **do**
13:         **if** Receive_Update_Request.$\Delta'[x]$ == TRUE **then** Update_Confidence.$\Delta'[x]$++
14:         **end if**
15:     **end while**
16:     ▷ Incorporate the update.
17:     **if** Update_Confidence.$\Delta'[x]$ ≥C$\Delta'[x]$ **then**
        Update_Local_Policy($\Delta'[x]$,Security_Ontology)
18:     **else**
        Discard($\Delta'[x]$)
19:     **end if**
20: **end if**

The described functions (As analysed in Sect. 5) either return a fixed-length array based on a parameters-request, or perform partial order sorting of a given fixed-length array. The only included loop is time dependent, corresponding to the data property T$\Delta'[x]$ that is also a fixed-length array predefined based on the lifetime of the altered property $\Delta[x]$. The value of T$\Delta'[x]$ monotonically decreases for every execution of the cycle. Hence, both algorithms terminate after the consecutive execution of the required steps.

Regarding the correctness of the algorithms, at the transmitter side applying the constraints identified at Sect. 4, allows for the consecutive execution of the required steps, for the update of the local policy and local audit mechanisms, in addition to the transmission of update requests to the required recipients in a prioritized order. At the receivers side with the reception of a non-incorporated alteration, the audit mechanisms are initially updated (AoRC), while the acceptance or rejection of the update request is based on a reconciliation confidence measure, calculated based on the number of estimated requesters. Given an update request, the local security mechanisms provide (through the node assignment list) a group of nodes that are co-located and serve the execution of the same action with the original update requester. Additionally, the local QoS mechanisms limit this list based on the current connectivity measures, providing the final *Requester_List'*, which includes the nodes that are expected and have the resources to transmit similar requests. If the sum of received requests meets the predefined criticality measure ($C\Delta'[x]$ can correspond to a percentage of *Requester_List'*), the alteration is accepted, incorporated and forwarder using the transmitter algorithm, otherwise it is recorded and rejected.

7   Conclusions

Through this article, the findings of our study regarding the reconciliation of ontologically defined security policies for tactical SOA, during the mission execution stage, have been presented. The primary contributions of this article are the investigation of the possible divergences and the identification of the required functionalities for policy reconciliation. The nature of occurring divergences has been limited to an expected and permitted subset, both in terms of scope, source and subject. Furthermore, the required functionalities for their reconciliation have been identified, taking under consideration the constraints of the tactical environment and the requirement for auditing, prioritization and roll back capabilities. Additionally, the developed mechanism for the consolidation of the reconciliation requirements and tactical constraints has been presented.

Our future plans include the further investigation and refinement of the proposed mechanism, with the incorporation of service invocation related metrics, and its extension within the scope of global or extended policy updates. Furthermore, within the scope of the ongoing project TACTICS, the analysis presented in this study and positive initial experimental results, are to be verified and demonstrated in large scale realistic scenarios.

Acknowledgments. The results described in this work were obtained as part of the EDA (European Defence Agency) project TACTICS (Tactical Service Oriented Architecture). The TACTICS project is jointly undertaken by Patria (FI), Thales Communications &Security (FR), Fraunhofer-Institut fur Kommunikation, Informationsverarbeitung und Ergonomie FKIE (DE), Thales Deutschland (DE), Leonardo (IT), Thales Italia (IT), Gjøvik University College (NO), ITTI (PL), Military Communication Institute (PL), and their partners, supported by the respective national Ministries of Defence under EDA Contract No. B 0980 GP.

References

1. Gkioulos, V., Wolthusen, S.D.: Securing tactical service oriented architectures. In: 2nd International Conference on Security of Smart Cities Industrial Control Systems and Communications-SSIC (2016)
2. Aloisio, A., Autili, M., D'Angelo, A., Viidanoja, A., Leguay, J., Ginzler, T., Lampe, T., Spagnolo, L., Wolthusen, S.D., Flizikowski, A., Sliwa, J.: TACTICS: tactical service oriented architecture. CoRR, vol. abs/1504.07578 (2015)
3. Lacy, L., Aviles, G., Fraser, K., Gerber, W., Mulvehill, A.M., Gaskill, R.: Experiences using OWL in military applications. In: Proceedings of the OWLED 2005 Workshop on OWL: Experiences and Directions, Galway, Ireland, November 11–12, 2005 (2005)

4. Semy, S.K., Pulvermacher, M.K., Obrst, L.J., Pulvermacher, M.K.: Toward the use of an upper ontology for U.S. government and U.S. military domains: an evaluation. Technical report, Submission to Workshop on Information Integration on the Web (IIWeb-04), in Conjunction with VLDB-2004 (2004)
5. Uszok, A., Bradshaw, J., Lott, J., Johnson, M., Breedy, M., Vignati, M., Whittaker, K., Jakubowski, K., Bowcock, J., Apgard, D.: Toward a flexible ontology-based policy approach for network operations using the kaos framework. In: Military Communications Conference, 2011 - MILCOM 2011, pp. 1108–1114, November 2011
6. Bunch, L., Bradshaw, J., Young, C.: Policy-governed information exchange in a U.S. army operational scenario. In: IEEE Workshop on Policies for Distributed Systems and Networks, 2008, POLICY 2008, pp. 243–244, June 2008
7. Lund, K., Eggen, A., Hadzic, D., Hafsoe, T., Johnsen, F.: Using web services to realize service oriented architecture in military communication networks. IEEE Commun. Mag. **45**, 47–53 (2007)
8. Trivellato, D., Zannone, N., Glaundrup, M., Skowronek, J., Etalle, P.S.: A semantic security framework for systems of systems. Int. J. Coop. Inf. Syst. **22**, 1–35 (2013)
9. Gkioulos, V., Wolthusen, S.D.: Enabling dynamic security policy evaluation for service-oriented architectures in tactical networks. Norw. Inf. Secur. Conf.-NISK **8**, 109–120 (2015)
10. Kolovski, V., Parsia, B., Katz, Y., Hendler, J.: Representing web service policies in OWL-DL. In: Gil, Y., Motta, E., Benjamins, V.R., Musen, M.A. (eds.) ISWC 2005. LNCS, vol. 3729, pp. 461–475. Springer, Heidelberg (2005)
11. Finin, T., Joshi, A., Kagal, L., Niu, J., Sandhu, R., Winsborough, W.H., Thuraisingham, B.: ROWLBAC - representing role based access control in OWL. In: Proceedings of the 13th Symposium on Access control Models and Technologie, Estes Park, Colorado, USA. ACM Press, June 2008
12. Blanco, C., Lasheras, J., Valencia-Garcia, R., Fernandez-Medina, E., Toval, A., Piattini, M.: A systematic review and comparison of security ontologies. In: Third International Conference on Availability, Reliability and Security, 2008, ARES 2008, pp. 813–820, March 2008
13. Souag, A., Salinesi, C., Comyn-Wattiau, I.: Ontologies for security requirements: a literature survey and classification. In: Bajec, M., Eder, J. (eds.) Advanced Information Systems Engineering Workshops. LNBIP, vol. 12, pp. 61–69. Springer, Heidelberg (2012)
14. Nguyen, V.: Ontologies and information systems: a literature survey. 6 (2011). http://digext6.defence.gov.au/dspace/handle/1947/10144
15. Gkioulos, V., Wolthusen, S.D.: Constraint analysis for security policy partitioning over tactical service oriented architectures. In: Advances in Networking Systems Architectures, Security, and Applications - of Springer's Advances in Intelligent Systems and Computing (2016)
16. Fudholi, D.H., Rahayu, W., Pardede, E.: A data-driven dynamic ontology. J. Inf. Sci. **41**, 383–398 (2015)
17. Zablith, F., Antoniou, G., d'Aquin, M., Flouris, G., Kondylakis, H., Motta, E., Plexousakis, D., Sabou, M.: Ontology evolution: a process-centric survey. Knowl. Eng. Rev. **30**(1), 45–75 (2015)
18. Besana, P., Robertson, D.: Probabilistic dialogue models for dynamic ontology mapping. In: Costa, P.C.G., d'Amato, C., Fanizzi, N., Laskey, K.B., Laskey, K.J., Lukasiewicz, T., Nickles, M., Pool, M. (eds.) URSW 2005 - 2007. LNCS (LNAI), vol. 5327, pp. 41–51. Springer, Heidelberg (2008)

19. Flouris, G., Plexousakis, D., Antoniou, G.: On applying the AGM theory to DLs and OWL. In: Gil, Y., Motta, E., Benjamins, V.R., Musen, M.A. (eds.) ISWC 2005. LNCS, vol. 3729, pp. 216–231. Springer, Heidelberg (2005)

20. Hooi, Y.K., Hassan, M.F., Shariff, A.M.: A survey on ontology mapping techniques. In: Obaidat, M.S. (ed.) Advanced in Computer Science and its Applications. LNEE, vol. 279, pp. 829–836. Springer, Heidelberg (2014)

21. Choi, N., Song, I.-Y., Han, H.: A survey on ontology mapping. SIGMOD Rec. **35**, 34–41 (2006)

22. Euzenat, J., Shvaiko, P.: Ontology Matching, 2nd edn. Springer, Heidelberg (2013)

23. Cobéna, G., Abdessalem, T., Hinnach, Y.: A comparative study of XML diff tools. Technical report, INRIA (2004)

24. Rana, V., Singh, G.: MBSOM: an agent based semantic ontology matching technique. In: 2015 International Conference on Futuristic Trends on Computational Analysis and Knowledge Management (ABLAZE), pp. 267–271, February 2015

25. Heflin, J. and Hendler, J. Dynamic ontologies on the web. In: Proceedings of the Seventeenth National Conference on Artificial Intelligence (AAAI-2000), pp. 443–449. AAAI/MIT Press, Menlo Park (2000)

26. dos Reis, J.C., Pruski, C., Reynaud-Delaître, C.: State-of-the-art on mapping maintenance and challenges towards a fully automatic approach. Expert Syst. Appl. **42**(3), 1465–1478 (2015)

27. Klein, M., Proefschrift, A., Christiaan, M., Klein, A., Akkermans, P.: Change management for distributed ontologies. Technical report (2004)

28. Bakillah, M., Liang, S.H., Zipf, A., Mostafavi, M.A.: A dynamic and context-aware semantic mediation service for discovering and fusion of heterogeneous sensor data. J. Spat. Inf. Sci. **2013**, 155–185 (2013)

29. Besana, P., Robertson, D.: How service choreography statistics reduce the ontology mapping problem. In: Aberer, K., et al. (eds.) ASWC 2007 and ISWC 2007. LNCS, vol. 4825, pp. 44–57. Springer, Heidelberg (2007)

30. Muthaiyah, S., Kerschberg, L.: Dynamic integration and semantic security policy ontology mapping for semantic web services (SWS). In: 2006 1st International Conference on Digital Information Management, pp. 116–120, December 2007

31. Khattak, A.M., Pervez, Z., Latif, K., Lee, S.: Short communication: time efficient reconciliation of mappings in dynamic web ontologies. Know.-Based Syst. **35**, 369–374 (2012)

32. Khattak, A., Pervez, Z., Khan, W., Khan, A., Latif, K., Lee, S.: Mapping evolution of dynamic web ontologies. Inf. Sci. **303**, 101–119 (2015)

33. Khattak, A., Latif, K., Khan, S., Ahmed, N.: Managing change history in web ontologies. In: Fourth International Conference on Semantics, Knowledge and Grid, 2008, SKG 2008, pp. 347–350, December 2008

34. Khattak, A.M., Latif, K., Lee, S.: Change management in evolving web ontologies. Know.-Based Syst. **37**, 1–18 (2013)

35. Stojanovic, L., Studer, R.: Methods and tools for ontology evolution. Technical report, Universitaet Karlsruhe (TH) (2004)

36. Benerecetti, M., Bouquet, P., Ghidini, C.: On the dimensions of context dependence: partiality, approximation, and perspective. In: Akman, V., Bouquet, P., Thomason, R.H., Young, R.A. (eds.) CONTEXT 2001. LNCS (LNAI), vol. 2116, pp. 59–72. Springer, Heidelberg (2001)

A Social Behavior Based Interest-Message Dissemination Approach in Delay Tolerant Networks

Tzu-Chieh Tsai[✉], Ho-Hsiang Chan, Chien Chun Han,
and Po-Chi Chen

Computer Science Department, National Chengchi University, Taipei, Taiwan
ttsai@cs.nccu.edu.tw,
{100753503,103753009}@nccu.edu.tw, hanjord@gmail.com

Abstract. Compared with 3G, 4G and Wi-Fi, Delay-Tolerant Networking (DTN) can only have intermittent chance to transmit messages. Without a clear end-to-end path, routing a message in DTN to the destination is difficult. But in some particular case, it could be an advantage. People around the world have their personal habit and it will be projected on their social life. Therefore we use the social behavior as a foundation feature of our routing algorithm. We propose two new kinds of routing algorithms with our own trace file. On one hand, birds of a feather flock together, so people who have similar interests tend to go to the same places. In case of this, we combining the personal interests and the trace file to different buildings where each node locates, we propose the building-based routing algorithm. On the other hand, we think people who have similar interests hang out together more often, so we use the social relationship as a feature and propose social-based routing algorithm. In the end, we compare our algorithms with Epidemic, MaxProp and PRoPHET routing algorithms. The result shows that our algorithms outperform the others.

Keywords: Delay-Tolerant network · DTN · Campus environment · Personal information · Personal interest · Social relationship

1 Introduction

1.1 Background and Motivation

In these years, smart phone has become more and more important in people's lives. Because smart phone is a very powerful device, it can be useful in many situations like sending and receiving e-mails, communicating with friends, acting as a digital calendar to remind us special days, checking daily weather, and virtual wallet. We can see everyone carry a smart phone wherever they go. Smart phone has become part of our lives.

The issue we focus on is plenty of advertising spam we might receive every time we open our email box. Maybe there are some messages we are interested in. However, we don't have time to check the spam one by one to pick what we want to read, so we ignore them in most cases. If messages can be transferred to people who are interested in them, it surely can reduce the overhead and make better performance. Transferring

R. Doss et al. (Eds.): FNSS 2016, CCIS 670, pp. 62–80, 2016.
DOI: 10.1007/978-3-319-48021-3_5

the messages through the Internet is not always the best way. First, transferring messages through the Internet is limited to its ability to access the Internet. Second, besides subscribing to specific channel s and receiving arbitrary spam, we can't get the messages we want. We want to push the messages to wherever the users are despite the ability to access the Internet. To overcome this problem, we think DTN (Delay-Tolerant Networking) is a good choice.

1.2 Delay-Tolerant Network (DTN)

Delay-Tolerant Networking is a dynamic wireless network. Every node may move freely and be organized depending on their social relationship. DTN can provide interoperable communication in challenging environments, which is defined as the network is not always in connect or the network has no end-to-end path. It is an approach of computer network architecture that can use the strategy of store, carry, and forward to transmit messages in the disconnected network environment. In Fig. 1, when the node is in the environment without network connection, it may convey messages to nearby nodes by using the short distance transmit technique such as Bluetooth or Wi-Fi direct. When the relay node receives the message, it can carry the message until meeting the next proper node to help transmit and forward the message. Via this approach, messages can travel around the environment and be transmitted to the destination. The connection between two nodes in the DTN environments can only keep for just few seconds, so it needs to find appropriate node to help transmit messages in limited time.

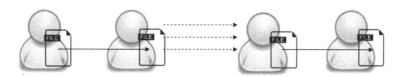

Fig. 1. Store, carry, and forward in DTN

There are two main reasons why we propose a routing algorithm based on DTN. First, we think that people have a routine trace everyday, just like most people have to go to work, and students have to go to school. We always get up around the same time in the morning, do the same chores, and most important of all, we commute to our destination in almost the same route and in regular time. It reveals that we might meet the same stranger every day, but we don't even notice. This stranger would be a terrific node in our routing algorithm. Because we can keep meeting this person everyday, we can update the messages with people we meet. Thus, we can know which person is closer to the destination that messages should be transferred to. Second, the Internet only provides us an end-to-end way to transfer the messages. Under this condition, if we want to transmit a message, we have to know where the destination is first. But in some cases, especially in advertising messages, we do not know where all the destinations are when the message is created. If we use the Internet as communication model, there are only three scenarios: (1) Enterprise, which creates the advertising messages, can only send these messages to people who have registered before. (2) People can only transfer messages to their friends. (3) The enterprise can spread the messages randomly, which will cost a lot. But in DTN, we can transfer the message through the node with store, carry and forward strategy. In the previous work [1], which is published in IEEE magazine, we can see that the research of using the interest as a feature in their routing protocol has an excellent performance. So we want to take it a step further and continue this research. In this paper, we suppose people who have similar interests tend to go to the same places. For example, people who like sports or exercise will go to the gym or sports field. Moreover, when they are shopping or doing something else, they are more likely to do things related to sports. People who are interested in art will go to see art exhibitions, and people who like reading will go to the library or bookstores. In this paper, we use this as a feature on our routing algorithm.

In tradition, there are several ways to route the message in DTN environment such as Epidemic, MaxProp [2], and PRoPHET [3]. The routing of Epidemic is by the way of transferring messages to every node where the carriers meet. The overhead of Epidemic is extremely high, but it has a better performance. We want to reduce the overhead as much as we can and prevent the performance from dropping too much. The routing of MaxProp concerns with people they have met before and the sequence of messages to be sent. MaxProp does not keep an eye on which direction the message should be sent to. In our routing algorithm, it will calculate the probability of which relay node gets a higher chance to arrive at the destination. PRoPHET focuses on two nodes meeting each other and which one gets a path that can transfer messages to the destination in a relatively higher probability. However, PRoPHET only works well at unicast case, and it saves all the probability of meeting all the other nodes, which would cost a lot of memory in a giant trace file. In our routing algorithm, we only saves data of the node that has met other nodes before, and it surely helps us reduce the usage of memory.

1.2.1 Cosine Similarity
In our routing algorithm, we have to decide whether people interest the advertisement message or not. For simplicity without losing of generality, we first choose cosine similarity as our indication for social interest relation. It is simple and quite easy to perform. We bring out the cosine similarity to help us determine who wants to know the message. The formula is showing below at Formula (1)

$$Cosine\ Similarity = \frac{\vec{A}, \vec{A_X}\,(D)}{\left\|\vec{A}\right\| \cdot \left\|\vec{A_X}\,(D)\right\|} = \frac{\sum_{i=1}^{n} A_i \times A_X(D)_i}{\sqrt{\sum_{i=1}^{n}(A_i)^2 \times \sum_{i=1}^{n}(A_X(D)_i)^2}}, i<n \quad (1)$$

In our trace file, we collect five different interests. But we only use two of the interests as the input. Because the rest three interest columns do not show the difference unfortunately. So, we only take two different interests as input column to calculate cosine similarity. The interest column will be assigned when a message is created. And all of the nodes in our simulator have their own interests. We can use both the interest column of the message and the interest column of the node to determine whether the node is interested or not. And the detail will be described in Sect. 4.

We propose a new DTN routing algorithm, based on the assumption that people who have daily routine and who have similar interests flock together. Then use the cosine similarity to see who is interested in the message. Finally, we compare our algorithm with the classic routing algorithm, and the result shows that we have a better performance.

2 Related Work

In the paper [4], the author thinks that social-based routing and location-based routing are slightly different. Due to the difficulty of collecting the real trace data, we can use the social data as a substitution. If we can collect both social-based and location-based data, we can compare them with each other.

In such a variety of DTN routing researches, we focus on two types of research (1) Collecting human real movement data. (2) How to use social data to send data to the destination quickly with less resource in DTN environment. Three trace data we will discuss below do not consider the condition of what the real society is like. There are different kinds of people with different interests in the real society, and this is what makes a diverse society. People may go to different places or do different things depending on their jobs and interests. So if we can use a more realistic trace file that we can regard it as a tiny real society. Different kinds of nodes are moving freely in the emulator, which is closer to the real society.

2.1 Social Trace Data

2.1.1 Reality Mining: MIT [5]

This experiment was carried out by MIT. The researcher gives 100 NOKIA's smart phone to 100 students, and the experiment duration is 9 months. Students who participated in the experiment were asked to use smart phones to communicate with other students by Bluetooth, and their trace, contact time, and communicate time were recorded. Via this experiment, we can analyze and predict social activities' relation with the subjects to know its next movement and social relations. The disadvantage of the experiment was that 75 students were from MIT Media Laboratory, and the other 25 students were from MIT Sloan business school. We think that the composition of participants can't be a miniature of the real society (Fig. 2).

2.1.2 Cambridge [6]

This experiment was carried out by Cambridge computer lab. The researcher used the equipment named iMote to collect the real trace data. In Cambridge05, the experiment separated students into freshman and sophomore, and it also included graduate and doctoral students (Fig. 3).

During the 11 days, 54 students used iMote equipment with Bluetooth technique, which helps to measure and record the main active area, the other students they contact,

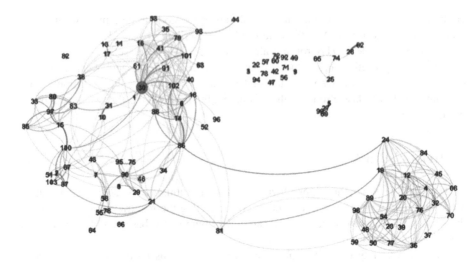

Fig. 2. One view of the network created by MIT reality mining dataset

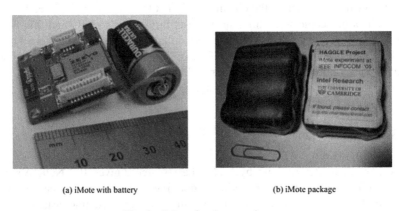

(a) iMote with battery   (b) iMote package

Fig. 3. iMote for the experiment

and the length of communicate time of each student. In the future, this data can be used in social relationship experiment.

2.1.3 Infocom05, 06 [7]
This experiment was held in students and professors who attended Infocom conference in 2005 and 2006. In the beginning, every participant was given an equipment named iMote. Because there were lots of different topics in the conference, every participant would go to listen to the topic they were interested in. Thus, we can know every participant's interests and who they communicate with.

In 4 days, 98 people participated in this experiment. Through the experiment, we can know each participant's professional specialty and whether they communicated with other people who have the same research domain., we can use participants' communicate time to conjecture their social relationship.

After reviewing previous research, we think we have to select the participants in order to make the trace data more similar to the real world. One of the most important things is to pick who can enroll in our experiment. All the details are described in Sect. 3.

In the trace file above, the participants are comprised of one or two particular group. Which will lead to the trace file is not general. And the trace file will be limited. What we want is a trace file that is a miniature of real society. So we will keep an eye on this while we are picking the participants to involve our experiment. And the detail will be described in Sect. 3.

2.2 Social-Based in Delay-Tolerant Network

2.2.1 Social-Aware Data Diffusion in Delay Tolerant MANETs [8]

This research proposes a routing algorithm based on different interests of each node. If two nodes have similar interests, which means the similarity of interests has exceeded the threshold, then we define them as friends; otherwise, the two nodes are defined as strangers. Therefore, when two nodes meet, they will exchange interest list and data list. When their interests are similar, they are friends to each other, and they will exchange data that the carrier likes. On the other hand, if they are strangers to each other, they will diffuse data that they are not interested in the message, just like the state shown below (Fig. 4).

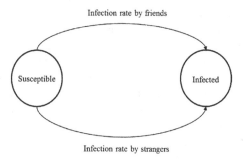

Fig. 4. Node infected by friend or stranger

Even if the two nodes are strangers, it doesn't mean that their friends or people they might meet are not interested in the message. So in our routing algorithm, we give a chance to case like that. We log every node that one node has met before, and we give it a try if the log file reveals they have a high chance to meet each other again and the data also interests them. For example, node A and node B go to work at the same time in the morning, and they take the same transportation. However, they are strangers, so they may not talk to each other. It is a good time to exchange the advertising messages they got while they are on the bus. During the time they are on the bus, they can know the interests of each other, and the interests of people who they usually meet. It helps us know whether other people are good relay nodes or not.

2.2.2   Social Network Analysis for Routing in Disconnected Delay-Tolerant MANETS [9]

It is difficult to diffuse messages in spare MANETs. All the nodes can move freely. So to find out the most efficient path is the key to this research. Previous research has conducted many theories to discover the best way to route the messages. To overcome this issue, there is Centrality, which can reveal whether the node is connected to the neighbor nodes. In other words, the node is able to know whether there is a path to the destination. They propose a new routing protocol named SimBet. SimBet is based on the betweenness Centrality and Similarity of nodes, and it chooses the intermediate node to help carry the messages. But the disadvantage is that all the messages tend to gather at some active nodes. Nodes, which are in relatively static state, can only transfer the messages and are not able to get the messages they want. In our routing algorithm, we do not want a few people to carry most of the messages. We think it might lead to information starving for some nodes. We consider not only the connection between nodes but also the landmark where the node has gone before. A node can be the relay node if this node will meet some other nodes which are interested in the message. Everyone can be the relay node depending on their movement and social relationship, and we think it is a better way to route the messages.

3   NCCU Trace Data

All the participants in trace data we mentioned above are limited to some specific group, either students of particular college or participants in particular conference. For example, the participants in MIT trace data were composed of Media Laboratory and business school students. In the Cambridge trace data, only computer laboratory students were enrolled. Furthermore, the Infocom trace data was collected during the conference, so most of the participants were related to the conference. We think these three trace data above can't reveal how the real social network works. In real social network, people are not supposed to do the same thing or the same work all the time, and they have different interests. We assume that where people usually go depends on their jobs and interests, and this issue is what we concentrate on. Because we don't have enough participants that can represent the real society to implement the experiment, we limit the environment to our campus. We recruit participators according to the ratio of different college. Also, the participators of the same college are from different departments for diversity. This is close to reality instead of participators with similar background like previous ones. Then, referring to [10], we not only record the trace file, but also record the self-declared interests of participators. In the end, because we don't have enough participants that can represent the real society to implement the experiment, we limit the environment to our campus. Participators are students in National ChengChi University.

3.1   Form (Selecting Participants)

When we were building our own trace file, we selected participants first. In order to find the suitable participants, we asked everyone who enrolled in to fill out the form, which asked some basic profile information and the most important thing, participants' college and interests. Figure 5 shows a part of the real data that we obtained.

ID	College	Sports	Reading	Social	Arts	Service
A	4	0.75	0.75	1	1	0.75
B	0	1	0.5	0.5	0	0.5
C	2	0.5	1	0.75	0.75	0.75

Fig. 5. Form list

3.1.1 College

All the participants should not come from one specific group, and they should not be all strangers to each other, either. Two participants were assigned to a group, and they must know each other or we would not accept them. We also cared about what college they are from. The college quantity ratio depended on the real college quantity ratio of total students in NCCU. If the quantity of one college was about to surpass the quantity we wanted, we would not accept the coming group, either. In this case, participants in our trace data were distributed to different colleges. Some people knew each other, and others didn't just like the real society.

3.1.2 Interest

In our algorithm, we had to determine where the destination of each message was. So we asked the participants what kind of interest they are. The interests were divided into five types, which were sports, reading, social, art, and service. The score was limited from 0 to 1, and each scale was at least 0.25. On the other side, when a message was created, it would be assigned to these five interest types accordingly. But when we calculating the average of these interest columns, we find out that the social, arts and service column don't show the difference. So we can only use sports and reading column as an input to calculate the cosine similarity as we mentioned earlier at Formula (1) to define whether the nodes are interested in the messages or not. If yes, the node will be one of the destinations of the message.

3.2 Trace Data

The most important part of trace data is the movement of each node. In order to get the real trace data, using smart phones was the best way to collect the information we needed. In traditional work, they created a device to record the user location, but the users may forget to bring the device with them. In reality, carrying another device only to send or receive messages is difficult to implement. In these days, people carry their own smart phone wherever they go, even at the places they are not allowed to use it. In conclusion, we think the smart phone is the best device to implement the DTN routing algorithm. For this reason, we designed an Android app and installed it in each participants' smart phone. We ran a background service to record GPS position every 10 min. If we scan too often, the battery of the smart phone will dry out fast. In our previous experiment, if we scanned every 5 min, the battery couldn't hold on for a day. To continue record the user position, we stipulated a scan every 10 min.

After all the trace data were collected, we ignored the trace data that were not on the campus and normalized the trace data. In addition, we want the trace data to move smoothly but not to disappear in one place and appear in another place suddenly. We had to normalize the trace data and keep the trace track continuous. Finally, when these works were done, our trace data were about to be used. There were 115 available data in our trace data in total, and the experiment lasted for two weeks, from 17th Dec to 31st Dec in 2014. All the trace data can be downloaded as soon as the paper is published [11].

4   Routing Approach

Most traditional routing scenarios only have one single destination, but our goal is to deliver messages to various destinations and reduce the overhead. However, using DTN to transmit a personal message to another person would not be as suitable as the advertising message. Thus, if a company wants to spread an advertising message to as many customers as possible, DTN is a suitable option.

4.1   Environment Definition

First, because the trace data we obtained were from the campus, we limited the simulation environment to the campus, too. When an advertising message is created, we don't quite know which destination we should send it to. All we can do is to find out the people who might be interested, and push the message to the right person. But in reality, we won't know whether a node is interested in the message or not until it bumps into other nodes and exchange metadata. The metadata includes the information about the contact node and other's interests, what nodes the contact node has met, and also what messages the contact node has got so far. When the receiver receives the metadata, it will calculate the cosine similarity between the interest type of the message and the interests of other nodes. If the result surpasses the threshold, then the message will be delivered. The entire process is termed "contact."

4.2   Routing Strategy

The message forwarding flow is shown in Fig. 6. Whenever a message is created, it will be assigned to one node randomly. For instance, we assume the message is assigned to node A. Node A may meet node B along the way to its destination. The time node A and node B spend on exchanging data with each other is called the total "contact time". When the connection is interrupted or out of the connection area, the data transferring stops. During the contact time, node A will try all kinds of messages

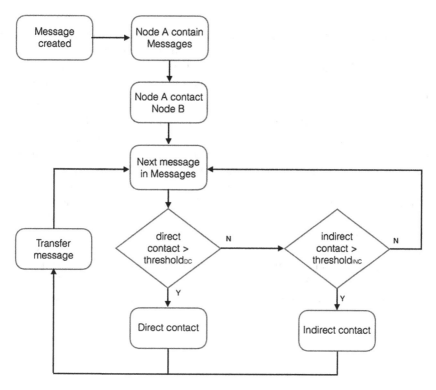

Fig. 6. Message forwarding flow

in the transferring waiting queue one by one. In next paragraph, there are two kinds of contact in our algorithm: direct contact and indirect contact.

4.2.1 Direct Contact

Direct contact is the most common way to transfer the messages. When two nodes meet each other, they will exchange metadata about their interests. And the node will calculate and compare the interest type of the message with the interests of another node it meets. For instance, node A encounters node B, and they exchange the metadata. Node A gets the interests of node B, and it will calculate the cosine similarity (Cos) between the interest of node B and the interest type of message (I_m). Node A will compare every message that node A has and node B doesn't have with the interest of node B (I_{N_B}). Formula (2) is shown below.

$$Cos(I_{N_B}, I_m) > Thres\_DC \qquad (2)$$

If the result of cosine similarity is greater than the *Thres_DC*, the message will be put in queue and ready to be transferred. The transferring time will last as long as the contact time of two nodes or until all the messages in queue are delivered.

4.2.2   Indirect Contact

In addition to direct contact, we expect the contact node can be a relay node, which can carry the messages to other nodes that are interested in the message. If we transfer these messages randomly, it will lead to high overhead and doesn't make our performance better. So we propose two kinds of indirect contact routing algorithms, which are base on building and social relationship.

Building Based Indirect Routing. We use the historical data of which building the students went to forecast the future. First, students of different colleges go to different buildings. For example, students of Commerce College have a higher chance to meet one another than any other student of other colleges because they usually go to the classrooms in the Commerce Building. Besides, students of Accounting Department have a much higher chance to meet students of the same department than students of Statistics Department in Commerce College. Second, we assume that students who have similar interests gather in the same building.

In Fig. 7, the students of Language College go to the library more often compared with the students of Science College and Law College. The reason might be that some ancient documents only have print editions, so most of them have to go to the library to get these books. However, the students of Science College can search the information on the Internet.

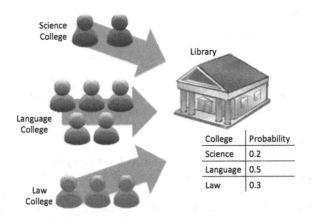

College	Probability
Science	0.2
Language	0.5
Law	0.3

Fig. 7. The counting of different college of students go to library

In order to prove our assumption, we calculate all the probability that students from different colleges go to each building on the campus. We can calculate the interests, which can be defined just like the form we ask the participants to fill out. Suppose that the interest of the building can be counted and defined according to what kinds of people have come before. Figure 8 displays the counting process. We first initialize all the interest of each building to 0. Second, whenever someone goes to the building, we add the interest of that person to the interest of the building. After the entire trace file is checked, we normalize the interest of the building. Finally, we can define the interest of the building.

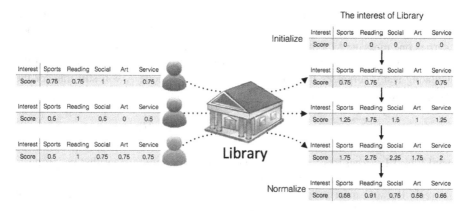

Fig. 8. The counting of the interest of library

When we finish collecting the entire trace file, we calculate how many times students of specific college go to specific building. We think there is logical evidence that students of the same college or the same department have a higher chance to go to the same building.

In our trace file there are 8 different colleges in total 115 nodes (S). And there are we define 17 buildings (B) to calculate the probability of college to each building $(Prob_{College(N_i),b_k})$ and try to verify our proposal. The calculating algorithm will be described below.

Algorithm1: Calculating probability of which building do students in college go
$Input: every\ node\ i(N_i) \in S, trace\ file(TF), every\ building\ b_k \in B$
$Output: Prob_{College(N_i),b_k}$
1: $Set\ every\ Prob_{College(N_i),b_k}\ to\ 0$
2: $For\ every\ N_i \in S$
3:   $For\ every\ time\ j(T_j)\ in\ TF$
4:    $if\ location\ (N_i, T_j)\ in\ b_k$
5:     $Prob_{College(N_i),b_k}(College(N_i), b_k) += 1$
6:   $end\ for$
7: $end\ for$
8: $Normalize\ Prob_{College(N_i),b_k}\ to\ between\ 0\ to\ 1, for\ every\ b_k$

Algorithm2: Calculating algorithm of the interests of building
$Input: every\ node\ i(N_i) \in S, Interest\ of\ N_i(I_{N_i}), trace\ file(TF), Interest\ of\ building\ b_k(I_{b_k})$
$Output: I_{b_k}$
1: $Set\ every\ I_{b_k}\ to\ 0, b_k \in B$
2: $For\ every\ N_i \in S$
3:   $For\ every\ time\ j(T_j)\ in\ TF$
4:    $if\ location\ (N_i, T_j)\ in\ b_k$
5:     $I_{b_k} = I_{b_k} + I$
6:   $end\ for$
7: $end\ for$
8: $Normalize\ I_{b_k}\ to\ between\ 0\ to\ 1, for\ every\ b_k$

After we do both Algorithms 1 and 2, we can use them to implement our Building-Based Indirect Routing. For instance, when node A encounters node B, node A carries one message M, but node B doesn't want message M. Then A will check the interest of every building and the probability of node B going to each building. If the calculating result surpasses the threshold (*Thres_meet*), node A will still transfer the message M to node B. The formula is shown in Formula (3).

$$\sum\nolimits_{b_k \in B} Cos(I_M, I_{b_k}) * Prob_{College(N_B), b_k} > Thres_{meet} \tag{3}$$

Social Based Indirect Routing. This is another indirect routing algorithm we propose. We think every one has a routine schedule in a period of time. In most cases, the period is a week. Because most people have to go to work or school on weekdays, it leads to the result that we will do mostly the same task like what we did seven days ago. In our campus scenario, students have to follow their own schedule and go to class accordingly. So if we can record where they were last week, we may forecast where they will go in the near future. To achieve our goal, we define a new interest (*IL*) that records what kind of people the node met seven days ago. The new interest of node A last week is defined as IL_{N_A, D_1}. The calculating algorithm of the new interest is described below.

Algorithm3: Calculating the new interest of node (*IL*)
Input: every node $i(N_i) \in S$, Interest of $N_i(I_{N_i})$, trace file(TF)
Output: IL
1: *Set every IL_{N_i, D_d} to $0, N_i \in S_N, D_d$ between Monday to Sunday, $1 \le d \le 14$*
2: *For every $N_i \in S$*
3:   *For every time $j(T_j)$ in TF*
4:     *if N_i meet N_x on $D_d, N_x \in S$*
5:       $IL_{N_i, D_d} = IL_{N_i, D_d} + I_{N_x}$
6:       $IL_{N_x, D_d} = IL_{N_x, D_d} + I_{N_i}$
7:   *end for*
8: *end for*
9: *For every $N_i \in S$*
10: *Normalize IL_{N_i, D_d} to between 0 to 1, for every $1 \le d \le 14$*

For example, node B encountered node C, D and E last Monday. So the new interest of node B (IL_{N_B, D_1}) will be the mean of the interests of node C, D and E, just like what we do when counting the interest of the building (Fig. 9).

When node A which has a message M meet node B next Monday, node B is not interested in the message M. Node A will check the IL_{N_B, D_1} value. If the cosine similarity between the message M and the new interest IL_{N_B, D_1} surpass the threshold (*Thres_Cos*). Node A will still transfer the message M to node B. Then node B will be

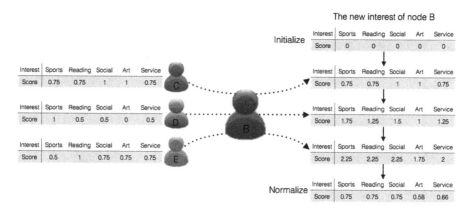

Fig. 9. The counting of the new interest of node B

a carrier of message M and help spread the message M. The formula is displayed below.

$$Cos\left(I_M, IL_{N_A,D_d}\right) > Thres_{Cos} \tag{4}$$

When we simulate our trace file, we only have two weeks of trace data. So, if we wt to simulate the first week, we have to use the IL_{N_i,D_d} of the second week. In other words, we have to use the $IL_{N_A,D_{d+7}}$ in our simulator if we want to simulate the first week. On the other hand, when we want to simulate the trace file of the second week, we use $IL_{N_A,D_{d-7}}$.

5 Simulation Settings

5.1 Simulation Environment

In our simulation, we use ONE (Opportunistic Network Environment simulator) [12] (shown as Fig. 10) and the map of NCCU (Nation Cheng-Chi University) surrounding area to validate our approach. All nodes in the simulation represent the student of this college, and they walk around according to their class schedule or for some purpose.

5.2 Simulation Setting

The simulation setting is shown as Table 1. The map area is 3764 m x 3420 m, which is the main active area of NCCU (Fig. 11), and the simulation time is from 12 a.m. to 12 a. m. of the next day. This is about 172800 s, it is equivalent to two days. The reason why we choose this time slot is that some students are at the school during the day and others live in the dorms at night. The node data transmission rate is 250KBps, and the transmission rage is 10 m. The message size of data is 500 KB ∼ 1 MB, the node buffer size is 100 MB, and the message's TTL is 1080 min, which is equivalent to 18 h.

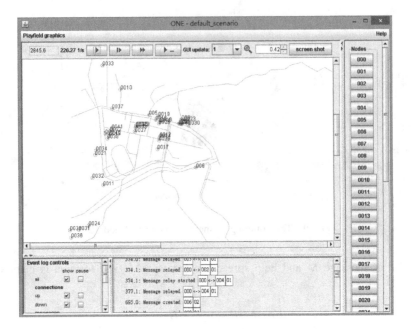

Fig. 10. One simulator

Table 1. Simulation settings

Area	6750*5100 m
Simulation time	172800 s
Data rate	250 KBps
Radio range	10 m
Message size	500 KB ~ 1 MB
Buffer size	100 MB
Total message created	62
Time to live	1080 min

5.3 Simulation Results

We simulated our trace file with both Building-based and Social-based routing algorithms, and we used the setting that we have described above. We evaluated the performance by checking the deliver rate and the overhead. If we can get higher deliver rate with lower overhead, it means that we get better performance. The deliver rate is what we cared about most. The results are as follows.

5.3.1 Delivery Ratio

Formula (5) is the way we counted the deliver rate. Because we had various destinations for one message (m_i), we had to know how many destinations the message would go to $(DestinationNum)$ and whether these destinations received the message

Fig. 11. NCCU surrounding area

(*DestinationRelayed*). The total message number (*TotalMessageNum*) created in one day was 62 messages. After we got the 2 parameters above, we could calculate the mean deliver rate of messages (*M*) per day. Finally, we could get the deliver rate.

$$Delivery\ Ratio = \frac{\sum_{m_i}^{M} \frac{DestinationRelayed}{DestinationNum}}{TotalMessageNum}, \forall m_i \in M \qquad (5)$$

Figure 12 displays the result that our routing algorithms have a better performance than MaxProp and PRoPHET, but they still have something to improve to reach the Epidemic.

In traditional simulation environment, the nodes can only be influenced by some particular factors like the nodes are in particular group, so there is a key routing feature that can be used. Some of the simulation results could reach the performance of Epidemic. But in reality, there are too many factors to affect people's behavior. For example, people may catch a cold, so they have to go to the hospital or clinic instead of working space. Furthermore, students may skip the class and go out to get some fun. Besides, we may run into somebody we know and then go to a coffee shop, which is not in our plan. So many factors can have an impact on our lives. It is hard to forecast the future precisely, and this is why we cannot beat the Epidemic routing algorithm.

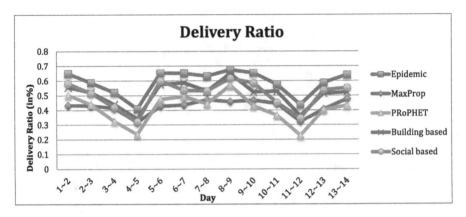

Fig. 12. Delivery ratio

5.3.2 Overhead

In order to have a better understanding, we calculated overhead with overhead copies and overhead ratio. The calculating formula is shown below. And we separate into two parts of overhead.

$$Overhead\ Copies = \frac{\sum_{m_i}^{M} Relayed - DestinationRelayed}{TotalMessageNum}, \forall m_i \in M \qquad (6)$$

$$Overhead\ Ratio = \frac{\sum_{m_i}^{M} \frac{Relayed - DestinationRelayed}{DestinationRelayed}}{TotalMessageNum}, \forall m_i \in M \qquad (7)$$

In Figs. 13 and 14, we can see that we have lower overhead than Epidemic, and we have a little bit lower overheads than MaxProp and PRoPHET relatively. Combined with Fig. 12 proves that the Building-Based and Social-Based routing algorithms have a better performance than the three traditional routing algorithms.

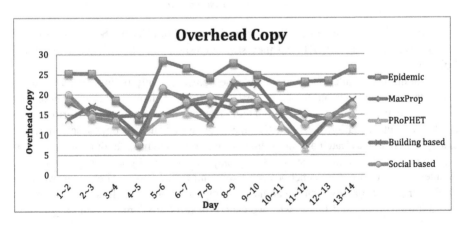

Fig. 13. Overhead copy

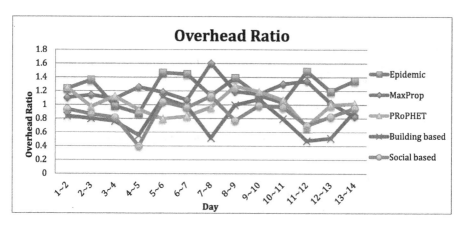

Fig. 14. Overhead ratio

6 Conclusion and Future Work

In this paper, we propose Building-Based and Social-Based routing algorithms. We use the trace file and the interests of people, which were collected in our experiment. Nodes in our experiment can not only be a destination node, but also be a relay node. It is helpful to spread messages. When two nodes meet, they will exchange metadata. We check whether the nodes are destinations and whether they are good relay nodes to make sure that the message transferring is efficient. Finally, we evaluate our routing algorithms with other algorithms, and the result shows that our algorithms have a better performance. In the future, we will consider using the real social relationship between each node. We think if two nodes know each other, they will have a higher probability to meet again. If this hypothesis is true, it will be a good feature to check whether the node is a good relay node or not.

References

1. Tsai, T.-C., Chan, H.-H.: NCCU trace: social-network-aware mobility trace. Commun. Mag. IEEE **53**, 144–149 (2015)
2. Burgess, J., Gallagher, B., Jensen, D., Levine, B.N.: MaxProp: routing for vehicle-based Disruption-Tolerant Networks. In: Proceedings of IEEE Infocom, April 2006
3. Lindgren, A., Doria, A., Schelen, O.: Probabilistic routing in intermittently connected networks. In: Dini, P., Lorenz, P., de Souza, J.N. (eds.) Service Assurance with Partial and Intermittent Resources (SAPIR 2004). LNCS, vol. 3126, pp. 239–254. Springer, New York (2004)
4. Zhu, K., Li, W., Fu, X.: Rethinking routing information in mobile social networks: Location-based or social-based? Elsevier Computer Communications (to appear), 2014
5. Eagle, N., Pentland, A.: Reality mining: sensing complex social systems. Pers. Ubiquit. Comput. **10**(4), 255–268 (2006)

6. Hui, P.: People are the network: experimental design and evaluation of social-based forwarding algorithms, Ph.D. Dissertation, UCAM-CL-TR-713. University of Cambridge, Computer Laboratory (2008)
7. Srinivasan, V., Motani, M., Ooi, W.T.: Analysis and implications of student contact patterns derived from campus schedules. In: Proceedings of ACM MobiCom, Los Angeles, CA, pp. 86–97 (2006)
8. Zhang, Y., Gao, W., Cao, G., Porta, T.L., Krishnamachari, B., Iyengar, A.: Social-aware data diffusion in delay tolerant MANET. In: Thai, M.T., Pardalos, P.M. (eds.) Handbook of Optimization in Complex Networks: Communication and Social Networks. LNCS, vol. 58, pp. 457–481. Springer, New York (2010)
9. Daly, E., Haahr, M.: Social network analysis for routing in disconnected delay-tolerant MANETs. In: Proceedings of ACM MobiHoc (2007)
10. Socievole, A., De Rango, F., Caputo, A.: Wireless contacts, Facebook friendships and interests: analysis of a multi-layer social network in an academic environment. In: 2014 IFIP Wireless Days (WD), IEEE (2014)
11. https://github.com/NCCU-MCLAB/NCCU-Trace-Data
12. Keränen, A., Ott, J., Kärkkäinen, T.: The ONE simulator for DTN protocol evaluation. In: Proceedings of the 2nd International Conference on Simulation Tools and Techniques, March 2009

The Looking-Glass System: A Unidirectional Network for Secure Data Transfer Using an Optic Medium

Gal Oren[1,2(✉)], Lior Amar[3], David Levy-Hevroni[2],
and Guy Malamud[2]

[1] Department of Computer Science,
Ben-Gurion University of the Negev, P.O.B. 653, Beersheba, Israel
`orenw@post.bgu.ac.il`
[2] Department of Physics, Nuclear Research Center-Negev,
P.O.B. 9001, Beersheba, Israel
`dlhevroni@gmail.com, guy.malamud@gmail.com`
[3] Parallel Machines - Information Technology and Services Ltd.,
Tel-Aviv, Israel
`liororama@gmail.com`

Abstract. The Looking-Glass system is a unidirectional network for data transfer using an optic medium, base on the principle of transferring information digitally between two stations without an electric connection. The implementation of this idea includes one side encoding and projecting the information to a screen in high speed, and a receiving side, which decodes the information image back to its original form. The decoding is done using a unique algorithm. Also, in order to synchronize between the transmitter and the receiver sides a separate synchronization system base on video pattern recognition is used. This technique can be useful whenever there is a need to transfer information from a closed network – especially sensitive one – to an open network, such as the Internet network, while keeping the information in its original form, and without any fear of an uncontrolled bidirectional flow of information – either by a leakage or a cyber attack.

Keywords: Confidential networks · Information security · Unidirectional systems · Encrypted data transmission

1 Introduction

"Oh, what fun it'll be, when they see me through the glass in here, and can't get at me!"

Lewis Carroll, *Through the Looking-Glass* (1871)

There are two major risks in the process of data transference from a secure confidential network into an open network [1]. The first risk is based on the reasonable assumption that there are possible threats which would like to break the networks' security fence in order to steal confidential information (in the best-case scenario), or to inject a malware in order to harm the network (in the worst-case). In the latter, one can

© Springer International Publishing AG 2016
R. Doss et al. (Eds.): FNSS 2016, CCIS 670, pp. 81–97, 2016.
DOI: 10.1007/978-3-319-48021-3_6

assume that a leak of secure information is possible due to metadata which was embedded into the transmitted data without the knowledge of the transmitter.

In many cases, the way those risks are handled is by an act of avoidance from any digital transmission of data from a secure network to an open network, when the alternative is by exporting a physical copy of the file by printing it on a plain paper. If it is necessary to get a digital copy of the data, the printed paper is scanned as an image, and by a usage of the Optical Character Recognition (OCR) technique it is been transformed back to a digital form in the open network [1]. This kind of solution holds many technical and immanent problems, which make the whole process ineffective at the best-case scenario because of its mechanical fashion, or even unreliable in the worst case because of the OCR algorithm inherent imperfect capabilities.

The Looking-Glass system supplies a technological solution to the current mode of work while supplying a secure solution to the two major risks of digital data transfer discussed above. The implementation of the idea includes a high speed transmitting side (i.e. transmitter), encoding and projecting the information to a screen, and a receiving side (i.e. receiver), which receive the information by remotely filming the screen and decoding the image information back to its original form, using unique image-processing algorithm. In order to synchronize between the transmitter and the receiver, we built a separate synchronization system based on video pattern recognition. Figure 1 shows a flow graph of the systems' activity, from the start of transmission until its end.

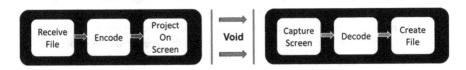

Fig. 1. The principles of information transport in the Looking-Glass system.

The system allows only a flow of ASCII characters from one network to the other without any physical channel that connects the two networks. This transference is done solely in a transfer only mode – meaning a 100 % secure system from any malware injection, simply because there is no bridge into it (in differ of the current solutions such as a unidirectional optical-fiber cable [2] or the Pump invention by the American Naval Forces [3]).

In addition, the fact that the system transfer ASCII characters explicitly means that there is no risk of a secure metadata hidden in the data file, simply because there are only characters and not a file. Also, in order to maintain a high reliability, several quality control features and optimizations were embedded in the system algorithm in order to prevent the inherent problems of unidirectional data transmit.

In this paper we discuss the Looking-Glass system and its algorithms. Section 2 will introduce the physical system and the principles of the development of the code written for the transmitting and receiving sides; Sect. 3 will elaborate and describe the encoding and decoding methods that were implemented at the transmitting and receiving sides, respectively; and at last, Sect. 4 will describe the system's benchmarks.

2 The Physical System

The physical infrastructure of the Looking-Glass system includes usage of two phys-ically separate computers, one as a transmitting unit and one as a receiving unit. On the transmitting side, a computer has been installed with two screens (24 inch, 1080p) – a screen for monitoring at the front and an internal screen inside the box to project the encoded data to the cameras at the receiving side. At the receiving side, a computer has been installed with one screen for monitoring at the front, and two cameras inside the box to film the transmission screen – an SLR (Single-Lens Reflex) camera and a valid Video camera (30 frames per second) to track and monitor the stream of transmitting frames. The units were separated 1-meter away from each other – more than the sufficient distance in order to capture all of the transmitting screen.

In order to prevent transmission of data using the magnetic field between the computer of the transmitting unit (which connects to the secured confidential network) and the computer of the receiving unit (which connects to the open network) [4–6], the two computers have been placed in two separated and sealed Faraday cages on a mobile facility, designed for the Looking-Glass system in a modular fashion base on ITEM profiles, 40×40 cm^2. The sealed box was chosen to maintain a constant quality

Description	System Sketch	System Pictures
The Looking-Glass system in closed mode. The system is not active at this mode.		
The Looking-Glass system in open mode. The system is active at this mode.		
An internal look into the Looking-Glass system.		

Fig. 2. Sketches and pictures of the Looking-Glass System.

of frames filming, without dependence on external lighting. This box can be opened and closed using tracks, and is installed on a mobile stand as described in Fig. 2.

The system code was developed in Python 2.7, using the OpenCV library [7], on Ubuntu operating system, and which all were installed on the computers.

2.1 The Transmitting Side

The implementation of the file transmission process is described in the algorithm below. In the first step, the data is divided into n equal segments of m characters in length, when the last segment will be equal to or less than the set size. The algorithm is performed $n + 1$ times, including one calibration segment. In every such action, one segment of the data is encoded, signed and presented to the screen.

Transmitting side – the algorithm for transmitting.

```
1. Divide the data to n segments of m characters.
2. Transmit the calibrated matrix of colors and
   shapes.
3. i = 1.
4. While i <= n:
   4.1   Encode segment i according to the scheme.
   4.2   Sign the segment.
   4.3   Export the following to the transmission
         monitor:
         4.3.1 The matrix of the encoded m characters.
         4.3.2 The signature embedded into a QR code.
         4.3.3 The mark of screen substitution or end of
               screen transmission.
         4.3.4 The textual information about the encoded
               segment.
   4.4   i = i+1.
```

Figure 3 shows a scheme of the transmitting monitor placed inside the box. In the top part, it is possible to see the m characters encoded into a matrix of squares in different colors. At the bottom of the figure, it is possible to see the control area, where the QR code is shown (which represents the signature of the encoded segment) and the geometric figure that is used as a feedback for frame transfer, or the starting and ending of a transmission (all steps will be explained next).

2.2 The Receiving Side

The receiving side includes two processes: filming and decoding. The filming process was implemented using a technique that is initialized automatically after initializing the transmission process. During the process execution, progress is reported to the monitoring screen.

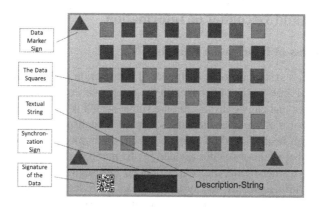

Fig. 3. The scheme of the transmitting side.

The algorithm developed for the receiving side is base on a usage of two cameras: an SLR camera, which performs the filming of the screen, and a Video camera, which continuously tracks the control area of the transmitting screen and provides a feedback to the filming process. The difference between the cameras is base on their quality of resolution: The video camera, which works in video mode, gives a low quality image, sufficient for real time decoding, while the SLR camera captures a high quality image, which is needed for the capture of large volumes of data, and for the quality assurance of the decoding algorithm. The frames filmed using the SLR camera are saved and decoded, while the images of the video camera are decoded but not saved; they are sampled through the entire duration of transmission until the end of the entire process, and are decoded solely in order to determine the status of the transmitted frame (substitution or end of transmission).

The identification mark chosen is a transfiguration of two geometric shapes, from a triangle to a square, which can be captured by the video camera. Using the change of shapes on the screen, a mark is given for filming and decoding of another image. Using another mark, a quadrilateral, it is possible to indicate that the process has reached its end (signaling the termination of the processes at the receiving side). The filming process is executed continuously as long as no last-image mark has been recognized by the Video camera, as shown in the algorithm below.

Receiving side – the algorithm for shooting pictures.

```
1. While the video camera hasn't recognized end-of-
   transmission configuration:
   1.1  Check the image transfer mark:
      1.1.1 If there was a shape transfiguration:
          1.1.1.1   Capture the transmission screen,
                  decode it, and save the results.
2. End process.
```

3 Encoding and Decoding of Information

3.1 The Process of Information Encoding

The Looking-Glass system encodes each byte of information using color squares, such that each byte of information is converted to a separate collection of squares. The number of squares that can be displayed on a single screen is a function of the squares total size and the size of the area where the squares can be displayed. Given the number of squares that can be displayed on a single screen, the number of displayed bytes per frame is the number of those squares divided by the number of squares per byte. Given the number of bytes per frame, the data is divided into segments of this length. After dividing the file into segments, the system encodes each segment into a sequence of squares. Each byte is encoded to a number of squares (base on the system configuration), the squares are presented as a matrix – from left to right and from top to bottom – and the color of each square is selected from a palette of defined colors whose size is the number of colors that the system is requires to differentiate between. There are two colors that are not part of the information encoding: black (0, 0, 0) and white (255, 255, 255). The black color is used for mark of boundaries, and the white color is used for background. An example of encoding the words "The Dog" can be shown in Fig. 4.

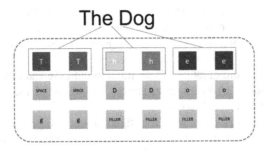

Fig. 4. Encoding the words "The Dog" into color squares (The letters are for demonstration).

The figure shows 3 lines of color squares, and every line has 6 squares. Each character is encoded using 2 squares, and the string "The Dog", which is composed of 7 characters (including one whitespace), is encoded to 14 information squares, and 4 additional squares are added as placeholders.

During the system development, we chose to code each byte of information using two color squares, such that each byte of information is converted to a separate collection of squares. These two squares provide a more than the sufficient amount of alphanumeric characters necessary for possible ASCII text file encoding. The squares color panel contains 16 colors that provide a sufficient distance in the color spectrum to successfully differentiate between the different colors in the decoding process. Therefore, in order to encode each character of the ASCII code, we use two squares of 16 different colors (16 × 16 combinations), where each square's color represents a number between 1 and 16. This encoding method was chosen to make the conversion process from image back to digital information as simple as possible. This format allowed representing an encoded data in size of 3–4 kB on the transmitting monitor, with 80 lines and 90 columns of squares. By that, one obtains a decoding ability at a very high level of reliability. Obviously, these encoding parameters can be vastly larger than in this prototype, and they are depends, as previously mentioned, on the SLR camera and the transmitting screen resolution.

The reason for choosing this encoding method rather than transmitting an image of the plain text is due to the following: if one had transmitted the information as raw data into the screen, and then try to decode it after filming from a significant distance (which creates the effects of filmography such as curvature and changing lighting), one wouldn't have obtained the high reliability we currently achieved using the squares encoding method, to which we made many image processing optimizations in order to adjust it to a remote screen filming (as will be explained in the hereby section). It is possible that a high reliability could have been obtained by filming the information in a raw form and decoding it using OCR algorithms, but only at the cost of enlarging the fonts sizes, which would have made the amount of information per transfer critically smaller – a state which eventually would cause to a significant slowdown of the whole system functionality.

3.2 Decoding the Information Area

After filming the transmitting screen, the receiving side decomposes the image structure into two parts and decodes each part separately. The parts are the information area and the control area, while the main phase is the decoding process of the former. This area consists of color squares, aligned line by line, meaning that by the end of the decoding process, we should obtain a list of all of the square's color indexes, when the order in this list is according to the order of appearance of the squares in the screen from left to right and from top to bottom. The decoding of the information area includes the following steps:

1. Converting the image to black-and-white scale and eliminating noise.
2. Accurately marking the squares matrix boundaries.
3. Identifying the squares correct order of appearance in every line.
4. Using the calculated squares position, average the real color for each square.

In the following sub-chapters we will review each of those steps and its algorithms.

3.2.1 Converting to Black-and-White Scale and Eliminating Noises

The image-processing algorithm that recognizes the squares locations needs an image where the squares are marked in black and the matrix spaces boundaries are marked in white. In order to obtain such an input image, we needed to convert the original image of the information area to solely shades of black-and-white scale, and to clean out all noises. The sub-steps in this process include converting the image to shades of gray; using a *Threshold* algorithm to convert the image to black and white where information squares are marked black and spaces are marked white; and cleaning noises, which are black spots that do not represent information squares. The first sub-step conversion process can be seen in the following comparison (Fig. 5).

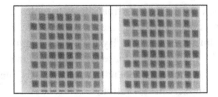

Fig. 5. The square matrix in color (left) vs. in shades of gray (right).

The next step is converting the gray-scale image into black-and-white scale image. In order to achieve that goal there is a need to use a *Threshold* algorithm [8], which can provide separation into black and white using a threshold value, where anything higher than this value is mapped to white, and anything lower than this value is mapped to black. However, when the information area is represented by a small image (a few hundred pixels in every direction), this technique works well, but when the image is large (as in our case), using a single threshold value to perform separation is not sufficient due to changes in the values that represent the spaces between the squares in the various areas of the image. For instance, in the shades-of-gray image in Fig. 5, the values of the pixels of the spaces is varying from values of 165–175 in one area of the image, and values of 170–180 in another area. In addition, we found that in occasion the values of color squares in a specific area are matching to the range of the space values in another area. The three following images (Fig. 6, from left to right) show the results while using the *Threshold* algorithm statically with values of 160, 170 and 180. It is possible to see that using a threshold value of 160 results in squares that completely deleted, when few squares attached to each other. It is also possible to see that although using a threshold value of 170 is not causing the information squares to disappear, it is causing them to become inseparable. Lastly, a usage of a threshold value of 180 shows that many squares turned into one black chunk. Therefore, we can conclude that it is not simple to find a sufficient threshold value which results a well defined black-and-white squares matrix.

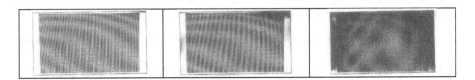

Fig. 6. Performing *Threshold* algorithm with values of 160 (left), 170 (middle) and 180 (right).

The solution for this problem is given by using the adaptive version of the *Threshold* algorithm. This version does not use a single value, but maps the area around a pixel in order to determine the threshold value. An example to this algorithm usage can be seen in the left side of Fig. 7, which shows an excellent separation of the information squares, without any disappearing squares and without inseparable squares zones. In order to calibrate the *Adaptive Threshold* function [8] we used empirical checks. Nevertheless, the image still contains black pixels originating from noise or environmental causes. The right side of Fig. 7 shows a zoom-in of the results after the *Adaptive Threshold*, and shows black pixels above the squares, which are not unified in large groups, and are not around the squares size.

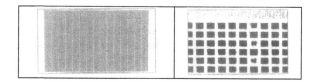

Fig. 7. The frame after using the *Adaptive Threshold* algorithm (left) and a zoom-in (right).

However, the next decoding algorithms still require a clean image. In order to clean the image, we will use two known filters – *Erode* and *Dilate* [8]. The *Dilate* function schematically calculates the maximal value for the area around the pixel, and changes the pixel value to this value. This operation lessens the black area, and hence makes single pixels, or a small amount of pixels combined together, disappear. The *Erode* function calculates the minimal value in the area around a pixel and changes the pixel's value to this value. This operation makes the black areas thicker, and hence thickens the square's boundaries that were earlier diminished by the *Dilate* operation. The three following images (Fig. 8, from left to right) respectively show the obtained noise after applying *Adaptive Threshold*; the noise elimination by *Dilate*; and thickening the squares back using *Erode*. At the end of the *Erode* operation, we obtain a clean image, on which the decoding process can be applied.

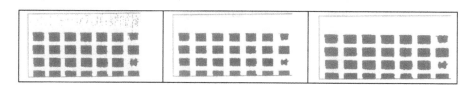

Fig. 8. Information squares with noise (left); after *Dilate* (middle); and after *Dilate* and *Erode* (right).

3.2.2   Marking the Matrix Boundaries

After converting the image using *Adaptive Threshold* and eliminating background noises, the obtained image contains the exact locations of most of the squares in a clear form. However, some of the squares, especially those containing light colors, are not completely hermetic and occasionally marked in more than one black area. In order to fix this situation – which does not enable correct recognition of the squares – we perform a process of identifying the separation lines (the matrix boundaries) between the different rows and columns of the squares matrix. The premise of this process is that even though there are squares that are not marked accurately, the most of the squares are intact and in a hermetic form, and they can be some good indicators to a total reconstruction of the original matrix boundaries. Using this knowledge, we can generate a new image where the squares are accurately defined. This algorithm is applied in the following form:

Marking the matrix boundaries algorithm.

```
1. Let SQ_ZONE, a matrix representing the image after its
   conversion    to    black-and-white-scale    and    after
   performing an eliminating noises process.
2. Let  GRID,  a  matrix  set  with  zeroes  (representing
   black) with the same dimensions of SQ_ZONE.
3. Let  N,  a  parameter  that  determines  the  number  of
   continuous  white  pixels  required  to  decide  if  a
   scanned segment is a matrix boundary.
4. For every line in SQ_ZONE, do:
   4.1.     Measure  the  longest  segment  of  continuous
      white pixels (CWP).
   4.2.     If running into a black pixel (square/noise):
      4.2.1.     Check whether CWP is approximate to N.
         4.2.1.1. If yes,  paint  the  CWP  segment  in  the
            GRID matrix in white, as a boundary.
         4.2.1.2. Otherwise, call 4.1.
```

The three following images (Fig. 9, from left to right) show the state of the image after the previous cleaning process; the result after scanning the rows; and the result after scanning the columns (by a 90 degrees' rotation for a scan as for rows) with overlap with the previous results. It is possible to see that the squares positions are marked in a hermetic form. It is clearly seen that some damaged squares boundaries, such as square [3, 7], were reconstructed hermetically.

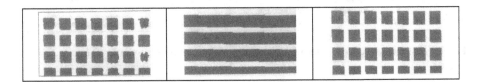

Fig. 9. The state of squares after the cleaning process (left); after scanning the rows (middle); and after scanning the columns with an overlap with the scanned rows (right).

Another example of this boundaries re-mark process can be seen in Fig. 10 which shows another area of the original image where the squares were marked inaccurately, and occasionally even discontinuously (meaning, with a few disconnected black spots on the same area of a square). It can be seen that the re-marking process had fixed the bisected squares, and that after the process they were marked accurately.

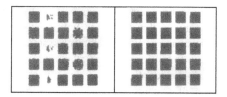

Fig. 10. Segment of information with faulted, split information squares (left) vs. the information squares after redraw (right).

3.2.3 Discovering Squares Correct Sequence by Walking-on-Line Algorithm

After resulting a black-and-white scale image with the squares appearing without noises and after a full marking process, the next step is to recognize the squares sequence and to store the squares in a data structure in the order they appear in the image. In order to do so, our Walking-On-Line algorithm is base on a simple fundamental which state that given a black pixel, it is possible to find the square shape boundaries that containing the pixel. This fundamental exists in the OpenCV library [5] and is implemented using the *FloodFill* algorithm. The basic idea underlying this square recognition algorithm is to start the scanning process from the top left corner, and scan from left to right and top to bottom. Each time the algorithm runs into a black pixel, the square containing the pixel and its connected pixels are recognized. Additionally, all of the pixels in the discovered square are set with a unique index, base on their location, in order to differentiate between the discovered squares and the squares that already been discovered. Using this method, when the next line of pixels is scanned and the algorithm runs into a pixel that belongs to an already-recognized square, the algorithm would skip it and not recognize the same square twice. In this recognition fashion, squares are found according to their order of appearance.

Seemingly, this algorithm does solve the problem; however, we note that the algorithm is based on the assumption that a line of squares is a straight line. This was found to be incorrect when filming long lines of squares. The camera lens creates a distortion, which makes the received image have iris curvature [9]. Additionally, the camera positioning might be un-aligned to the filmed screen perfectly, which makes the lines appear tilted. An example for this curvature can be seen in Fig. 11. The figure shows a green line starting at the left part of the second line of the squares. It can be seen that this line (a straight, balanced line) reaches the upper pixels of the first lines in the image shown, meaning that some squares of the first line will be recognized only after some information squares of the second line have already been recognized – a situation which cannot be tolerable.

Fig. 11. The curvature in the information squares is highlighted by the balanced green line. (Color figure online)

The difficulty introduced by lines curvature can be dealt by assuming that the height difference between two squares within the same line is negligible, e.g. two or three pixels between each square at most. For the current system design, this assumption enables recognition of the information squares adaptively at every image decoding (since the curvature is not always identical). By using this method, the problem can be overcome. The recognition of squares sequence is performed using the algorithm below.

Walking-on-Line Algorithm.

```
1. Scanning the pixels in the image from left to right,
   top to bottom.
   1.1.     When running into a black pixel:
      1.1.1.       Calculate the boundaries of the square
                   that contains the pixel, compute its center,
                   and save the results.
      1.1.2.       Walk right from the calculated center
                   of the square until running into a black pixel
                   which not belongs to the current identified
                   square, or until reaching the end of the line.
         1.1.2.1. If a black pixel found:
            1.1.2.1.1. Call 1.1.1.
      1.1.3.       If reaching the right end of the line:
         1.1.3.1. Continue to the next line.
         1.1.3.2. Call 1.1.
2. When there are no more pixels to scan:
   2.1.     End process.
```

Figure 12 shows an example of the algorithm actions in order to recognize the squares correct sequence. First, the top left pixel of the first square in the line is recognized (marked in red). The arrows show the process of recognizing the sequence, starting from the recognized pixel, towards the center of the square, and finally to the right, until a new square is recognized. By using this method, the heights of the scanned squares are continuously fixed, and the squares are captured based on their order of appearance. At the end of the recognition algorithm's run, all of the squares are obtained and ordered in the order of their appearance, and they are finally ready to be decoded back to the ASCII characters they represent.

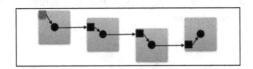

Fig. 12. Recognizing the squares correct sequence in an adaptive scanning. (Color figure online)

3.2.4   The Color Decoder and Its Calibration

After achieving to get the correct boundaries of the squares and its correct sequence, the remaining step is to convert each original color of each square into the number the color represents, and then back to its ASCII code. As previously explained, every character – with 8-bit limit in this prototype – is encoded into two squares, meaning every square represents four bits (which represent 16 values). Therefore, 16 different colors are in need, when every color of every square represents an index between 1 and 16. The task for the color decoder is to convert a color sample in RGB format from the correctly marked square to an index between 1 and 16.

However, after filming the screen, the obtained color for each square was found to be not identical to the color the square originally painted in at the encoding side. Furthermore, it has been found that some colors located in different areas of the transmitting screen were decoded with different coloring, even though the squares were originally painted in the exact same color. This phenomenon is caused by lighting conditions, which can be different in each part of the screen, as well as optic distortions that affect the color sample. For example, a color square that was originally painted at the encoding side with the color [240, 163, 255] was sampled at the decoding side with three different values of [140, 106, 205], [144, 109, 210], and [145, 111, 211] at three different zones of the frame. Nevertheless, based on the measurements it seems that there is no great spatial proximity between the sampled values; however, these values – as well as the rest of the samples – are relatively close to each other on the color spectrum, meaning that the distortion in the conversion of the original color does not provide a great scattering of the values.

Either way, the color decoder needs to take this phenomenon into account and overcome it. This is done by calibrating the decoder by color 'zones', and not by a 'one-to-one' value, meaning that although a color can be decoded with a deviation from the original RGB representation, the algorithm still will be able to recognize that a distorted RGB representation represents a specific color which have an ASCII index translation. In order to achieve this goal, the transmitting side encodes a calibration image as the first frame, containing squares which represent all of the 16 different colors in all of the different parts of the frame, and which the decoding algorithm at the receiving side knows all of its precise original RGB representations and indexes, base on its locations. Then, in order to represent all of the original indexes in all of the possible zones, a 3-dimentional matrix of $256 \times 256 \times 256$ values – which represents all of the RGB spectrum – is set with the original indexes at each of the RGB representations, including all of its nearest neighbors. For example, if it is known that the green color index was set to be 7, and if it was found that the decoding of the color square which represented this color at a specific zone in the frame was a RGB representation of (76, 153, 0), then the value of those indexes and of all of its nearest neighbors will be set to be 7 in the 'translation' matrix for further decoding of the colors of the squares (exemplification in Fig. 13).

Note that since the chosen 16 colors are relatively far from each other on the spectrum, it is reasonable to assume that there will be no overlap between the squares' RGB representation in the three-dimensional matrix.

The Color Detector and its Calibration Algorithm.

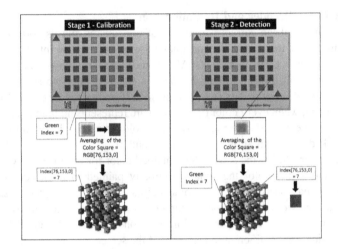

Fig. 13. Example for square decoding: Calibrating the green color (on the left) and using the calibration matrix to decode it (on the right). (Color figure online)

1. Set the matrix ORIG_CALIBRATION_MAT with the original color indexes of the calibration matrix.
2. For every square in the calibration decoded frame (DECODE_CALIBRATION_MAT):
   2.1.   Calculate the square average color using RGB fashion, such that the R represents a X coordinate index, the G represents a Y coordinate index, and the B represents a Z coordinate index of a 3-dimentional matrix (CALC_CALIBRATION_MAT) of 256X256X256 (the RGB spectrum).
   2.2.   Set the index of the original color from ORIG_CALIBRATION_MAT in CALC_CALIBRATION_MAT base on the R, G, and B indexes which calculated in 2.1 from DECODE_CALIBRATION_MAT, as well as in all of its nearest neighbors in the 3-dimensional matrix (a total of 27 squares that will have the same index).
3. For every square in the next decoded frames:
   3.1.   Calculate the square average color using RGB fashion.
   3.2.   Access CALC_CALIBRATION_MAT using the RGB indexes and check the color index which was previously set.
      3.2.1.   If index found:
      3.2.1.1. Return index.
      3.2.2.   Else, if no color index has been found:
      3.2.2.1. Start a rotational scan using the *Nearest Neighbor* technique until finding the color's matching index.

3.2.5   Decoding the Control Area and Quality Control

The decoding process of the control area (Fig. 14) consist of three different, independent parts:

Fig. 14. The information area. From right to left: the frame's title, the recognition square for frame switch, and the QR encoding square.

1. The first part focuses on the square and rectangle, which appears alternately. These shapes alternation indicate the receiving side on a change of the encoded information, meaning a new information transmission has been done, and there is a need to capture the new frame. A termination of the process is identified when a quadrilateral is being detected.
2. The second part focuses on the quality control aspect of the process, and includes decoding a QR code. As known [10], the QR code represents an accurate collection of characters (in this case, 256 or 512 characters, depending on the size of the square and the type of encoding). Using this knowledge, the encoding side takes the textual information which needs to be transmitted as a matrix of squares and uses a cryptographic hash function to obtain a checksum of this information, base on *sha256-512*. This checksum is encoded by the program into the QR square. Using OpenCV functions [7], this part of the control area is identified and decoded at the receiving side. When the receiving side has both the Checksum of the encoded information and the information itself as decoded by the decoding algorithm, the quality control algorithm encodes the information using the same cryptographic hash function and compares it to the checksum extracted from the QR square. If the comparison of the two checksums is found to be identical, it means that the information transfer has been completed successfully. Otherwise, it means that at least one character was missed or transferred incorrectly. This method ensures the completeness of the information transfer in an 100 % certainty (assuming checksum reliability).
3. The third part is an accurate documentation of the frame's name. This information is derived and saved in the folder the frame is saved in for logging purposes.

4   Performance Tests

Performance tests for the Looking-Glass system have been executed for three different stages of the system operation: the encoding stage, the transmission and receiving stage, and for the decoding stage. The distribution of time for a standard 3 kB frame in the system is shown in the following three steps. The speed of encoding, transmitting and receiving, and decoding of a single frame is currently stand on ~ 3 s.

1. **The encoding step:** the data is processed into squares which instantly displayed on the screen. This step is relatively faster than the other steps, and currently stand on less than a second.
2. **The transmission step:** the data is transmitted from the transmitting side to the receiving one. This step completely depends on the rate of the filming process (i.e. cameras properties) and the rate of the saving process (i.e. receiving computer properties), and it is currently stand on less than a second.
3. **The decoding step:** the data is extracted from the received frame, and a verification process confirms that all of the data that have been transformed back to the original digital form successfully. This whole process currently stands on less than two seconds (and this time factor can be significantly improved in the future using parallel processing at the decode process).

The three steps have been tested separately in 20 experiments. In order to verify the system's stability and reproducibility factors, a script was written to generate 20 simple, textual data files with random content, with sizes that varied from 10 kB to 100 kB, and from 100 kB to 1 MB – sizes which represents the common range of data sizes the system will need to transfer. The files were transferred continuously, and benchmarks was measured for each data transfer. It was found that the Looking-Glass system performances was linear to the amount of submitted data, as expected.

5   Conclusions and Future Work

The work presented in this paper lead to the conclusion that it would be beneficial to use the Looking-Glass system in order to securely and reliably transfer information digitally using an optic medium. Also, this paper opens a number of prospective directions for future research, which one immediate direction is to explore how to optimize the system infrastructure using new hardware, and how to increase the performances of the suggested algorithms, mainly using parallel processing.

Acknowledgments. This work was supported by the Lynn and William Frankel Center for Computer Science.

References

1. Shabtai, A., Elovici, Y., Rokach, L.: A Survey of Data Leakage Detection and Prevention Solutions. Springer Science & Business Media, New York (2012)
2. Okhravi, H., Sheldon, F.T.: Data diodes in support of trustworthy cyber infrastructure. In: Proceedings of the Sixth Annual Workshop on Cyber Security and Information Intelligence Research, p. 23. ACM, April 2010
3. Kang, M.H., Moskowitz, I.S., Chincheck, S.: The pump: A decade of covert fun. In: Computer Security Applications Conference, 21st Annual, p. 7. IEEE, December 2005

4. Kuhn, M.G., Anderson, R.J.: Hidden data transmission using electromagnetic emanations. In: Kuhn, M.G., Anderson, R.J. (eds.) Information Hiding. LNCS, vol. 1525, pp. 124–142. Springer, Heidelberg (1998)
5. Kramer, F.D., Starr, S.H.: Cyberpower and National Security. Potomac Books Inc, Lincoln (2009)
6. Zhao, N., et al.: EMI Spy: harnessing electromagnetic interference for low-cost, rapid prototyping of proxemic interaction. In: 2015 IEEE 12th International Conference on Wearable and Implantable Body Sensor Networks (BSN), IEEE (2015)
7. Suarez, O.D., Carrobles, M.D.M.F., Enano, N.V., García, G.B., Gracia, I.S., Incertis, J.A.P., Tercero, J.S.: OpenCV Essentials. Packt Publishing Ltd., Mumbai (2014)
8. Petrou, M., Petrou, C.: Image Processing: The Fundamentals. Wiley, New York (2010)
9. Goldberg, N.: Camera Technology: the Dark Side of the Lens. Academic Press, Boston (1992)
10. Furht, B. (ed.): Handbook of Augmented Reality. Springer Science & Business Media, New York (2011)

Privacy Preserving Consensus-Based Economic Dispatch in Smart Grid Systems

Avikarsha Mandal[(✉)]

Offenburg University of Applied Sciences, Offenburg, Germany
`avikarsha.mandal@hs-offenburg.de`

Abstract. Economic dispatch is a well-known optimization problem in smart grid systems which aims at minimizing the total cost of power generation among generation units while maintaining some system constraints. Recently, some distributed consensus-based approaches have been proposed to replace traditional centralized calculation. However, existing approaches fail to protect privacy of individual units like cost function parameters, generator constraints, output power levels, etc. In this paper, we show an attack against an existing consensus-based economic dispatch algorithm from [16] assuming semi-honest non-colluding adversaries. Then we propose a simple solution by combining a secure sum protocol and the consensus-based economic dispatch algorithm that guarantees data privacy under the same attacker model. Our Privacy Preserving Economic Dispatch (PPED) protocol is information-theoretically secure.

Keywords: Privacy · Economic load dispatch · Critical infrastructure protection · Consensus algorithm · Smart grid · Secure multi-party computation

1 Introduction

Economic dispatch problem (EDP) has been an important research topic to the power grid community over the past few decades [1,5,8,17]. The solution to this optimization problem is a power output combination of all generators in the grid which gives minimum total operating cost while maintaining several system constraints. The optimization techniques used in EDP may vary with various requirements. While numerical methods like lambda-iteration or gradient search [15] are conventionally used, use of more expensive techniques like particle swarm optimization [8] and genetic algorithms [1] can be found in the literature as well. Traditionally, the EDP calculation is done with a centralized control scheme. Here, one of the generators or a third party acts as a trusted leader. The leader collects different parameters and associated constraints from all generators. Next, it performs optimization calculations and sends the optimal solution to every generator. However, these centralized schemes fail to achieve requirements of the modern smart grid which promises a more distributed and reliable infrastructure. A small change in the smart grid or failure of the trusted leader requires a complete

© Springer International Publishing AG 2016
R. Doss et al. (Eds.): FNSS 2016, CCIS 670, pp. 98–110, 2016.
DOI: 10.1007/978-3-319-48021-3_7

change in the whole centralized scheme. Recently, researchers are applying distributed consensus algorithms to solve economic dispatch to overcome the challenges of centralized schemes [16,17]. In general, these algorithms are iterative and incremental cost (IC) is used as a consensus variable. While these proposals focused on convergence analysis while satisfying different system constraints, protecting the privacy of individual units was overlooked. For example, the cost function parameters of individual units are privacy sensitive. If these parameters get revealed to a competitor, it may try to reduce its own operational cost to establish itself as the cheapest utility provider in the market.

Our Contributions. Our contributions in this paper are fourfold. First, we identify which information should be private in EDP calculation. Second, we analyse the security of an existing distributed consensus algorithm [16] and show how different privacy sensitive data are leaking even under a simple semi-honest attacker model without any collusion. Third, we provide a solution to this problem by adding a privacy layer with a secure sum protocol. Fourth, we propose an information-theoretic privacy model for this type of protocols and give a security proof for the proposed protocol in our model.

The rest of the paper is organized as follows: In Sect. 2, related research work has been highlighted and Sect. 3 gives some background on economic dispatch and privacy sensitive data. In Sect. 4, we describe a non-private incremental cost consensus algorithm from [16] and show an attack with a single non-colluding semi-honest node. Sections 5 and 6 provide our proposed privacy preserving protocol and security analysis respectively. Finally, we conclude along with possible future work in Sect. 7.

2   Related Works

Although some distributed consensus algorithms for EDP are discussed in the literature, we are not aware of any studies which analyse privacy leakage or any protocol with a concrete security proof for EDP solutions. Some privacy-preserving attempts to generalized distributed iterative computation can be found in [3,9,14]. Different privacy preserving techniques based on homomorphic encryption [12], secret sharing [11] etc. can be used in the smart grid data aggregation. For more details, an overview of privacy preserving data aggregation techniques in the smart grid is presented in [7].

In [17], the authors adopted the equal incremental cost (IC) optimization criterion with lambda-iteration method and the IC of each generator is chosen as the consensus variable. IC is basically the increase in total cost resulting from an increase in power generation. The proof of the equal incremental cost criterion (i.e. when each generator has the same IC values, we have an optimal solution of EDP) can be found in [15]. In [17], each generator sends its own IC to its neighbor and the proposed consensus algorithm drives all individual IC to a common value. Importantly, the mismatch between demand and total power generated is fed back to the consensus algorithm to meet the demand constraint. However, the algorithm is not completely distributed because a leader has to be deployed to collect the power generated by each generator to calculate the total mismatch.

In [6], the authors took a different approach to solve EDP. They considered the total power generated by all generators as a linear piecewise continuous function of IC. In their decentralized approach, EDP is solvable with a ratio consensus algorithm if the demand lies in one of the linear segments. However, the ratio consensus algorithm is applied to learn the consensus parameters and generator constraints.

Yang et al. in [16] use similar IC criteria as in [17]. The authors in [16] considered a strongly connected network and claimed that in comparison to [17], in their algorithm every generator does not need to know the cost function parameters of the other generators. In this algorithm, the equal IC and estimation of mismatch between demand and total power are obtained collectively through local interaction between the generators. Moreover, their solution is distributed as no leader agent is needed to collect all the power generated by every generator. In Sect. 4, we show how privacy sensitive data can get leaked in the scheme from [16].

Some information-theoretically secure summation protocols against a semi-honest adversary can be found in [2,4,10]. The basic idea behind the secure sum protocol proposed in [2] is breaking a node's input in n pieces and distributing $n-1$ parts to other $n-1$ nodes in a full n node mesh network. An improved version of the protocol is proposed by Chor and Kushilevitz [4]. The protocol described by Kreitz et al. [10] is similar to [4] but can be practical for large networks.

3   Preliminaries

3.1   Economic Dispatch Problem

In this section, we give a brief background on the economic dispatch problem (EDP). The notation used in the paper is given in Table 1. Let us consider an n node generator system $(1, 2, \ldots, n)$ where $C_i(x_i)$ is the cost function of output power x_i. We can formulate the total cost of operation as:

$$C_{total} = \sum_{i=1}^{n} C_i(x_i) \qquad (1)$$

The objective of the economic dispatch problem is finding optimal values for all x_i's at which C_{total} is minimum. This cost function C_i is different for different nodes as they have different parameters. Furthermore, this minimization problem has to be solvable under some demand and generator constraints.

Under ideal conditions, the total power demand D should be equal to the total power generation (neglecting the transmission loss). Hence, the demand constraint equation can be written as:

$$D - \sum_{i=1}^{n} x_i = 0 \qquad (2)$$

Table 1. Nomenclature

Notation	Description
G	Network graph
V, E	Set of vertices and edges in G
n	Total number of generator nodes in the network G
i, j	Different generator nodes in network
t	Discrete time step for each round
$x_i(t)$	Output power of node i at round t
C_i	Cost function of node i
$\lambda_i(t)$	Incremental cost of node i at time step t
a_i, b_i, c_i	Cost function parameters
$\alpha_i, \beta_i, \gamma_i$	Cost function parameters used in [16]
D_i	Local power demand for node i
D	Total power demand in the network $(D = \sum_{i=1}^{n} D_i)$
$\underline{x_i}, \overline{x_i}$	Minimum and maximum output power limit of generator i
$y_i(t)$	Power mismatch of node i at round t
p_{ij}, q_{ij}	Different elements of admittance matrix P and Q of the network
N_i^+, N_i^-	In and out neighbours of node i
ϵ	Very small public constant

Furthermore, the following generator constraint equation has to hold:

$$\underline{x_i} \leq x_i \leq \overline{x_i} \tag{3}$$

The EDP is solvable when both Eqs. (2) and (3) satisfy.

The cost function is a quadratic function and can be formulated as:

$$C_i(x_i) = a_i x_i^2 + b_i x_i + c_i \tag{4}$$

Now, incremental cost is basically the increase in total cost resulting from an increase in power generation.

$$\lambda_i = \frac{dC_i(x_i)}{dx_i} = 2a_i x_i + b_i \tag{5}$$

The *equal incremental cost criterion* [15] met when every node has equal incremental cost, resulting in an optimal solution for the economic dispatch problem.

3.2 Privacy Sensitive Data

One should correctly identify which information can be privacy sensitive in the EDP calculation before designing a privacy preserving protocol. In the smart grid, power generator's output power, cost function parameters, and minimum/maximum output power are privacy sensitive data.

- **Output Power** (x_i): Revealing individual output power to other generators may harm the business model of the utility company. In a competitive market, when the power output of a generator is revealed to other competitors (i.e. other generators), they can try to generate more power during peak hour than that individual to oust him from the energy market.
- **Cost Function Parameters** (a_i, b_i, c_i): The cost function parameters of the individual generator are privacy sensitive information. The cost function is a critical business information to reveal to other generators in the smart grid. Knowing the operational cost of another individual, a competitor will try to reduce its own operational cost and establish himself as the least-cost utility provider in the market. Revealing the parameters (a_i, b_i, c_i) of the ith generator, a competitor will know the cost function of i.
- **Minimum** ($\underline{x_i}$)/ **Maximum** ($\overline{x_i}$) **Power Output:** Individual generator constraints are private sensitive. A generator might want to keep its generating capacity private from other generators. A competitor can manipulate its power generation based on other generator's output capacity and the demand curve to be the key player in the market.

4   Attack on Consensus-Based Solution from [16]

The existing consensus-based solutions in [16,17] use similar structures. In general, these algorithms are iterative and incremental cost is used as a consensus variable. We analyse the state of the art distributed solution for EDP proposed by Yang et al. [16] in this section. We believe the attack scenario will be similar for other existing EDP solutions.

4.1   System Model

In [16], authors assumed a strongly connected network topology as a directed graph $G = (V, E)$. The set of vertices $V = \{1, 2, \ldots, n\}$ represent the generator nodes of the network and the set of edges $E \subseteq V \times V$ represent the communication structure between the nodes. Strongly connected means, there exists a path between any pair of two vertices in the directed graph. A direct edge from i to j is denoted by an ordered pair $(i, j) \in E$ and means that a node j can receive information from i. The in-neighbors and out-neighbors of ith node are represented by $N_i^+ = \{j \in V | (j, i) \in E\}$ and $N_i^- = \{j \in V | (i, j) \in E\}$ respectively. A node can receive information from in-neighbors and send information to out-neighbors. As each node can know its own state information, each vertex belongs to both its in-neighbors and out-neighbors ($i \in N_i^+$ and $i \in N_i^-$). Note that this implies $\forall i \in V, (i, i) \in E$.

4.2   Description of the Incremental Cost Consensus Algorithm

Here, two matrices are defined as $P, Q \in \mathbb{R}^{n*n}$ based on the topology of the graph G. Let, p_{ij} and q_{ij} be the elements of matrices P and Q respectively. All the elements of P and Q are public. They are defined as:

$$p_{ij} = \begin{cases} \frac{1}{|N_i^+|} & \text{if } j \in N_i^+ \\ 0 & \text{otherwise} \end{cases}$$

$$q_{ij} = \begin{cases} \frac{1}{|N_i^-|} & \text{if } i \in N_j^- \\ 0 & \text{otherwise} \end{cases}$$

The standard cost function used for EDP calculation is quadratic as shown in (4). In [16], a slightly different quadratic cost function is being used (quadratic convex).

$$C_i(x_i) = \frac{(x_i - \alpha_i)^2}{2\beta_i} + \gamma_i \tag{6}$$

Where $\alpha_i \leq 0$, $\beta_i > 0$ and $\gamma_i \leq 0$. However, the cost functions (4) and (6) are basically equivalent if we replace $\alpha_i = -(b_i)/(2a_i)$, $\beta_i = 1/2a_i$ and $\gamma_i = c_i - (b_i^2)/(4a_i)$. /The IC of node i at a discrete time index t can be formulated as:

$$\lambda_i(t) = \frac{x_i(t) - \alpha_i}{\beta_i}$$

- **Initialization:** D_i is the local demand associated with the node i. Hence, the total demand is $D = \sum_{i \in V} D_i$.

In the initialization $\forall i \in V$, we have:

$$x_i(0) = \begin{cases} \overline{x_i}, & \text{if } \overline{x_i} < D_i \\ D_i, & \text{if } \underline{x_i} \leq D_i \leq \overline{x_i} \\ \underline{x_i}, & \text{if } D_i < \underline{x_i} \end{cases}$$

$$\lambda_i(0) = \frac{x_i(0) - \alpha_i}{\beta_i}$$

$$y_i(0) = D_i - x_i(0)$$

- **Main Algorithm:**

$$\lambda_i(t+1) = \sum_{j \in N_i^+} p_{ij}\lambda_j(t) + \epsilon y_i(t)$$

$$x_i(t+1) = \beta_i\lambda_i(t+1) + \alpha_i$$

$$y_i(t+1) = \sum_{j \in N_i^+} q_{ij}y_j(t) - (x_i(t+1) - x_i(t))$$

At each round, every node updates its own IC $\lambda_i(t)$ from the IC values of the previous round received from its in-neighbors N_i^+. Then, each node updates its output power $x_i(t)$ from the updated IC. Subsequently, i calculates the mismatch $y_i(t)$ and forwards to its out-neighbors N_i^- for the next round. ϵ is a positive scalar and controls the convergence speed.

4.3   Attacker Model

We assume the attacker i is a non-colluding semi-honest node in the network, i.e. i strictly follows the protocol but it may analyse the messages exchanged during the execution of the protocol to gain additional information. We also assume that the attacker knows the local demand D_j of any other node j in the network. This assumption is realistic as the local demand D_j is basically the aggregated demand from the consumers of j and can be a public information. An attack will be successful if the attacker achieves full knowledge about output power, generator constraints, and any cost function parameter.

4.4   Privacy Sensitive Data Leakage

Let us assume that the messages are exchanged between two nodes, i (attacker) and j (another node) where $i \in N_j^-$ and $N_j^+ \subseteq N_i^+$.

- **Output Power (x_j):** At $t = 0$, i receives $\lambda_j(0)$ and $y_j(0)$. As i knows the local power demand D_j, i can simply get the value of $x_j(0)$. Then, i can find out the output power x_j at every time step as i can get $\sum_{k \in N_j^+} q_{jk} y_k(t)$ from each k in N_j^+ at round t and $y_j(t+1)$ from j at round $t+1$:

$$x_j(t+1) = \sum_{k \in N_j^+} q_{jk} y_k(t) - y_j(t+1) + x_j(t)$$

- **Cost Function Parameter (a_j, b_j):** i knows values of $\lambda_j(t)$ and $x_j(t)$, from two iterations:

$$x_j(t) = \beta_j \lambda_j(t) + \alpha_j$$
$$x_j(t+1) = \beta_j \lambda_j(t+1) + \alpha_j$$

Solving two linearly independent equations, node i can find the values of α_j and β_j which will give values of a_j and b_j.
- **Minimum $(\underline{x_j})$ and Maximum $(\overline{x_j})$ Power Output:** At $t = 0$, if i observes $x_j(0) \neq D_j \neq 0$, i can get one of the values of $\underline{x_j}$ or $\overline{x_j}$.

5   Privacy Preserving Economic Dispatch (PPED) Protocol

To prevent the privacy leakage, we present our PPED protocol which uses a privacy layer in each round using a secure sum protocol. We can use the consensus-based algorithm proposed by Yang et al. [16] described in Sect. 4.2 as the basis. The secure sum protocol used in PPED is similar to the $n - private$ protocol for summation presented by Benaloh in [2] but uses $(n - 1)$ partitions instead of n.

5.1   System Model

We consider a complete synchronous network $G = (V, E)$ with n nodes and a secure and reliable point to point communication channel between every node (no eavesdropping). As G is a complete graph, every element in P and Q matrix will be $\frac{1}{n}$.

A practical assumption would be that all numerical values we want to compute are fixed point values for EDP algorithms. We can multiply any fixed point elements with a suitable constant and convert them into integers. In Step 2 we convert $\lambda_i(t)$ and $y_i(t)$ to integer and in Step 7 we convert these values back to the fixed point domain.

5.2   PPED Protocol

The diagram of our PPED protocol with 4 nodes is illustrated in Fig. 1. The protocol is outlined as follows:

- **Step 1**: Initialization of every node at $t = 0$:

$$x_i(0) = \begin{cases} \overline{x_i}, & \text{if } \overline{x_i} < D_i \\ D_i, & \text{if } \underline{x_i} \leq D_i \leq \overline{x_i} \\ \underline{x_i}, & \text{if } D_i < \underline{x_i} \quad \forall i \in V \end{cases}$$

$$\lambda_i(0) = \frac{x_i(0) - \alpha_i}{\beta_i}$$

$$y_i(0) = D_i - x_i(0)$$

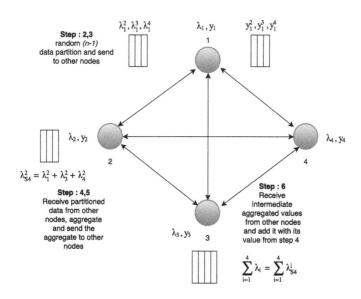

Fig. 1. PPED protocol for $n = 4$

- **Step 2**: In step 2 to step 6 we map $\lambda_i(t)$ and $y_i(t)$ values to integer values and map it back in step 7. We consider $\mathbb{Z}_M$ as the additive group of integers from 0 to $M-1$ (M is a large integer, $M > \sum_{1=1}^{n} \lambda_i(t)$ and $M > \sum_{1=1}^{n} y_i(t)$). Every node i, chooses $(n-1)$ numbers independently with uniform distribution from $\mathbb{Z}_M$, such that their sum (in modulo M) is equal to $\lambda_i(t)$. The same is done for breaking $y_i(t)$ into $n-1$ parts where each segment is allotted for a specific node. For example, i creates the segment $y_i^{(j)}(t)$ for the node j. All calculations are done in modulo M.

$$\lambda_i(t) = \sum_{\forall j \in V, j \neq i} \lambda_i^{(j)}(t)$$

$$y_i(t) = \sum_{\forall j \in V, j \neq i} y_i^{(j)}(t)$$

- **Step 3**: Every node i sends distinctive segments $\lambda_i^{(j)}(t)$ and $y_i^{(j)}(t)$ to the respective j in the network. We can use some token to distinguish between $\lambda_i(t)$ and $y_i(t)$ segment values.
- **Step 4**: Each node i will receive a total of $n-1$ data segments for $\lambda(t)$ and $y(t)$ respectively from the other $n-1$ nodes in the graph. All the nodes add all their received segments for $\lambda(t)$ and $y(t)$ separately. Let us consider at i, with total received sums $\lambda_{S4}^i(t)$ and $y_{S4}^i(t)$ respectively.

$$\lambda_{S4}^i(t) = \sum_{\forall j \in V, j \neq i} \lambda_j^{(i)}(t)$$

$$y_{S4}^i(t) = \sum_{\forall j \in V, j \neq i} y_j^{(i)}(t)$$

- **Step 5**: Every node i sends $\lambda_{S4}^i(t)$ and $y_{S4}^i(t)$ to the remaining $n-1$ nodes in the network.
- **Step 6**: Every node i adds all received $\lambda_{S4}^j(t)$'s from the other $n-1$ nodes and its own $\lambda_{S4}^i(t)$. Hence, every node gets sum value without knowing the individual inputs:

$$\sum_{\forall i \in V} \lambda_i(t) = \sum_{\forall j \in V, j \neq i} \lambda_{S4}^j(t) + \lambda_{S4}^i(t)$$

Do the same for $y(t)$.
- **Step 7**: Node i finds:

$$\lambda_i(t+1) = \frac{1}{n} \sum_{\forall i \in V} \lambda_i(t) + \epsilon y_i(t)$$

- **Step 8**: For all nodes i, the power output is calculated as:

$$x_i(t+1) = \beta_i \lambda_i(t+1) + \alpha_i$$

- **Step 9**: For all nodes i, the power difference is compared as:

$$y_i(t+1) = \frac{1}{n} \sum_{\forall i \in V} y_i(t) - (x_i(t+1) - x_i(t))$$

- **Step 10**: Check:
  if

$$\forall i \in V, y_i(t+1) \approx 0$$

  then EDP solution found **break**

  else t = t+1, Repeat $\forall i \in V$ Step 2 to Step 10.

6   Security Analysis

We consider a non-colluding semi-honest adversary for the security analysis. The privacy model proposed in this section is information-theoretic and inspired from [4,13]. To formally analyse the security of PPED protocol, we define correctness and privacy in our model. We start with the following notation:

- Let Γ be a randomized protocol which computes the optimal solution of the economic dispatch problem.
- At the beginning of Γ, each node i has 5 private inputs a_i, b_i, c_i, $\underline{x_i}$ and $\overline{x_i}$ without any probability space associated with them. We can represent private inputs of all nodes as Z, a $5 \times n$ matrix where i^{th} column denotes $\overrightarrow{z_i} = (a_i, b_i, c_i, \underline{x_i}, \overline{x_i})$ and 1st row denotes $\overrightarrow{z_a} = (a_1, a_2, \ldots, a_{n-1})$.
- $view_i^\Gamma$ is the set of information received by the ith node during an execution of the protocol Γ.

Then protocol Γ under a non-colluding semi-honest adversary should satisfy the following two properties:

Definition 1. *(Correctness) Each node is guaranteed to obtain its correct output. In this case, no attacker can alter the output of the computation i.e. each node is guaranteed to compute correct $\min_{x_i} C_{total}$ at the end of the protocol.*

Definition 2. *(Privacy) None of the nodes should learn anything more than what follows from its input and allowed output from the execution of the protocol. A non-colluding semi-honest adversary i does not learn any additional information from the execution of Γ if the following holds:*
For any two input matrices Z and W, which find the same output $(output_i^\Gamma(Z) = output_i^\Gamma(W))$ and agree with i such that $\overrightarrow{z_i} = \overrightarrow{w_i}$, the probability distributions of the set of information received by node i are equal i.e. each information received does not depend on any input. Hence,

$$\Pr(view_i^\Gamma | Z) = \Pr(view_i^\Gamma | W)$$

$output_i^\Gamma$ consists of $\sum_{\forall i \in V} \lambda_i(t)$ and $\sum_{\forall i \in V} y_i(t)$ for $t = \{0, 1, \dots\}$. This security definition guarantees in the worst case that any adversary only gains knowledge of $output_i^\Gamma$. This is not the strongest security notion, where any adversarial node i would only gain knowledge of the optimal power output of node i. However, even with the knowledge of $output_i^\Gamma$ or $\sum_{\forall i \in V} \lambda_i(t)$ and $\sum_{\forall i \in V} y_i(t)$ for $t = \{0, 1, \dots\}$, the adversarial node i can not gain any information about $(a_j, b_j, c_j, \overline{x_j}, \underline{x_j})$ or the optimal x_j for $j \neq i$. Note that even if the attacker node i knows D_j for some other node j, the attacker can only gain knowledge that $y_j(0)$ might be zero (if $\underline{x_j} \leq D_j \leq \overline{x_j}$ then $x_j(0) = D_j$) and that $\lambda_j(1)$ might be $\sum_{k \in V} \lambda_k(0)$ but everything else remains private as individual $\lambda_j(t)$ and $y_j(t)$ values are unknown.

Theorem 1. *(Correctness of PPED) If non-private economic dispatch algorithm in PPED is correct, the PPED protocol is correct.*

Proof. Correctness of non-private economic algorithm of PPED is proved in [16]. Hence the correctness of PPED protocol follows from, for any round for $t = \{0, 1, \dots\}$, $\sum_{\forall i \in V} \lambda_i(t) = \sum_{\forall j \in V, j \neq i} \lambda_{S4}^j(t) + \lambda_{S4}^i(t)$ and $\sum_{\forall i \in V} y_i(t) = \sum_{\forall j \in V, j \neq i} y_{S4}^j(t) + y_{S4}^i(t)$. □

Theorem 2. *(PPED protocol is private against any non-colluding semi-honest adversary i if $n \geq 4$)*

Proof. The view of node j consists of $(\lambda_i^{(j)}(t), \lambda_{S_4}^i(t), y_i^{(j)}(t), y_{S_4}^i(t))$, for $i \in V, i \neq j, t \in \{0, 1, \dots\}$. For node i, all of these are uniformly distributed with the constraints,

$$\sum_{i \in V} \lambda_i(t) = \sum_{i \in V, i \neq j} \lambda_i^{(j)}(t) + \sum_{i \in V, i \neq j} \lambda_{S_4}^i(t)$$

$$\sum_{i \in V} y_i(t) = \sum_{i \in V, i \neq j} y_i^{(j)}(t) + \sum_{i \in V, i \neq j} y_{S_4}^i(t)$$

This shows that the view of node i only depends on $output_i^\Gamma$, not on input matrices Z or W, unless $|V| = n \leq 3$ (in which case node j can recover values of $\lambda_i(t)$ for $i \neq j$). □

6.1 Communication Complexity

Now, let us see how much communication overhead is produced by our PPED protocol compared to a non-private economic dispatch protocol. A non-private protocol finds the solution by sending $2tn(n-1)$ messages, whereas the PPED protocol takes $4tn(n-1)$. In PPED, for step 3 there are $n(n-1)$ messages sent for $\sum_{\forall i \in V} \lambda_i(t)$ and $n(n-1)$ messages sent for $\sum_{\forall i \in V} y(t)$. In step 5, another $2n(n-1)$ messages are sent. Hence, the total number of communication messages is $4tn(n-1)$ for a t round PPED protocol. In terms of order of complexity, both protocols have $\mathcal{O}(tn^2)$.

7 Future Works

This work assumes a fully connected bidirectional topology. It would be interesting to extend this work to a relaxed topological constraint. Also, a stronger security notion with colluding adversaries is a future direction. Our assumption for the cost function here is quadratic as per the standard used by power engineers. However, the cost function could behave differently with different requirements in the smart grid. In the future, one can analyse the security of EDP solutions when the cost function changes in a particular setting.

Acknowledgments. The author would like to thank Erik Zenner and Frederik Armknecht for the helpful discussions on security analysis.

References

1. Bakirtzis, A., Petridis, V., Kazarlis, S.: Genetic algorithm solution to the economic dispatch problem. IEE Proc.-Gener. Transm. Distrib. **141**(4), 377–382 (1994)
2. Benaloh, J.C.: Secret sharing homomorphisms: keeping shares of a secret secret. In: Odlyzko, A.M. (ed.) Advances in Cryptology, Crypto 86. LNCS, vol. 263, pp. 251–260. Springer, Heidelberg (1986)
3. Bickson, D., Dolev, D., Bezman, G., Pinkas, B.: Peer-to-peer secure multi-party numerical computation. In: 2008 Eighth International Conference on Peer-to-Peer Computing, pp. 257–266. IEEE (2008)
4. Chor, B., Kushilevitz, E.: A communication-privacy tradeoff for modular addition. Inf. Proces. Lett. **45**(4), 205–210 (1993)
5. Chowdhury, B.H., Rahman, S.: A review of recent advances in economic dispatch. IEEE Trans. Power Syst. **5**(4), 1248–1259 (1990)
6. Dominguez-Garcia, A.D., Cady, S.T., Hadjicostis, C.N.: Decentralized optimal dispatch of distributed energy resources. In: 2012 IEEE 51st Annual Conference on Decision and Control (CDC), pp. 3688–3693. IEEE (2012)
7. Erkin, Z., Troncoso-Pastoriza, J.R., Lagendijk, R.L., Perez-Gonzalez, F.: Privacy-preserving data aggregation in smart metering systems: an overview. IEEE Signal Process. Mag. **30**(2), 75–86 (2013)
8. Gaing, Z.-L.: Particle swarm optimization to solving the economic dispatch considering the generator constraints. IEEE Trans. Power Syst. **18**(3), 1187–1195 (2003)
9. Huang, Z., Mitra, S., Vaidya, N.: Differentially private distributed optimization. In: Proceedings of the 2015 International Conference on Distributed Computing and Networking, p. 4. ACM (2015)
10. Kreitz, G., Dam, M., Wikström, D.: Practical private information aggregation in large networks. In: Aura, T., Järvinen, K., Nyberg, K. (eds.) NordSec 2010. LNCS, vol. 7127, pp. 89–103. Springer, Heidelberg (2012)
11. Kursawe, K., Danezis, G., Kohlweiss, M.: Privacy-friendly aggregation for the smart-grid. In: Fischer-Hübner, S., Hopper, N. (eds.) PETS 2011. LNCS, vol. 6794, pp. 175–191. Springer, Heidelberg (2011)
12. Li, F., Luo, B., Liu, P.: Secure information aggregation for smart grids using homomorphic encryption. In: 2010 First IEEE International Conference on Smart Grid Communications (SmartGridComm), pp. 327–332. IEEE (2010)
13. Lindell, Y., Pinkas, B.: Secure multiparty computation for privacy-preserving data mining. J. Priv. Confidentiality **1**(1), 5 (2009)

14. Naranjo, J.A.M., Casado, L.G., Jelasity, M.: Asynchronous privacy-preserving iterative computation on peer-to-peer networks. Computing **94**(8–10), 763–782 (2012)
15. Wood, A.J., Wollenberg, B.F.: Power Generation Operation and Control. A Wiley-Interscience publication. Wiley, Hoboken (1996)
16. Yang, S., Tan, S., Jian-Xin, X.: Consensus based approach for economic dispatch problem in a smart grid. IEEE Trans. Power Syst. **28**(4), 4416–4426 (2013)
17. Zhang, Z., Chow, M.-Y.: Convergence analysis of the incremental cost consensus algorithm under different communication network topologies in a smart grid. IEEE Trans. Power Syst. **27**(4), 1761–1768 (2012)

Network DDoS Layer 3/4/7 Mitigation via Dynamic Web Redirection

Todd Booth and Karl Andersson[✉]

Division of Computer Science, Luleå University of Technology,
97187 Luleå, Sweden
PhD@ToddBooth.Com, Karl.Andersson@Ltu.Se
http://OrcId.Org/0000-0003-0593-1253,
http://OrcId.Org/0000-0003-0244-3561

Abstract. Layer 3, 4 and 7 DDoS attacks are common and very difficult to defend against. The academic community has published hundreds of well thought out algorithms, which require changes in computer networking equipment, to better detect and mitigate these attacks. The problem with these solutions, is that they require computer networking manufacturers to make changes to their hardware and/or software. On the other hand, with our solution, absolutely no hardware or software changes are required. We only require the use of BGP4 Flow-Spec, which has already been widely deployed many years ago. Further the customers' own ISP does not require Flow-Spec. Our algorithm protects groups of over sixty-five thousand different customers, via the aggregation into one very small Flow-Spec rule. In this paper, we propose our novel, low cost and efficient solution, to both detect and greatly mitigate any and all types of L347 DDoS Web attacks.

Keywords: DDoS · DRDoS · Bandwidth · Reflector · BotNet · BGP4 · Flow-Spec

1 Introduction

Various acronyms and terms used in this paper, are defined in the Table 1. There are numerous academic papers, which provide the background, present case studies, and/or perform a literature survey concerning detecting and/or mitigating network based distributed denial of service (DDoS) attacks [2,6,9–11,17–21]. Therefore, this paper will limit the background and will not repeat the same numerous figures. As an illustration, in this paper we will refer to a Bank, as the on-line web service under attack. However, there is nothing bank specific in the solution, so it is applicable to any public Web service. Note that this paper is only a conceptual design and the experiment has been left as recommended future work.

© Springer International Publishing AG 2016
R. Doss et al. (Eds.): FNSS 2016, CCIS 670, pp. 111–125, 2016.
DOI: 10.1007/978-3-319-48021-3_8

Table 1. Acronym and term definition table

Term	Definition
BGP4	Border Gateway Protocol version 4
BotNet	A network collection of zombies (PCs infected with Malware)
CAPTCHA	Completely Automated Public Turing Test
CDN	Content Delivery Network
DoS	Denial of Service attack
DDoS	Distributed Denial of Service attack
DSR	Design Science Research methodology
Booters	DDoS attacks as a service (for rent)
DRDoS	Distributed Reflection DoS
IETF	Internet Engineering Task Force
IP	Focus in this paper is IPv4
ISP	Internet Service Provider
L4	Layer 4 (transport)
L7	Layer 7 (application)
L347	IP layer 3, 4 and/or 7 attacks
MPLS	Multi-protocol Label Switching
NATO	North Atlantic Treaty Organization
Null-Route	ISP basically discards all traffic, concerning the DDoS
NTP	Network Time Protocol
RFC	IETF request for comments document
SDN	Software Defined Networks
WAF	Web Application Firewall
Zombies	A collection of malware infected, remote controlled hosts

1.1   Research Problem

Information Systems, often include Web servers, which are accessible via the Internet (publicly facing). These Information Systems are being constantly being successfully attacked via network based DDoS attacks. There are a wide variety of network based DDoS attacks, such as network layer 3 (L3), transport layer 4 (L4) and application layer 7 (L7) attacks. Collectively, we will refer to these as L347 network attacks. A recent 2016 DDoS Internet attack measured over 400 Gbps [8]. Very close to 0 % of organizations have 400 Gbps, of ISP bandwidth, so these attacks cannot be prevented, by trying to only stop the attack, at the organizations' premises. So organizations sometimes try to have their ISP or Web server provider mitigate these attacks. However, many ISPs and cloud provides will not have enough free bandwidth to handle an attack of 400 Gbps. Even if they did, the solution to process the 400 Gbps stream is often very expensive. What many ISPs and cloud providers will do, is during a DDoS attack, they will null-route the organizations incoming traffic, until the DDoS is over, which means that the organization will be completely down. There are many other types of L347 network attacks. There are some really great solutions, however these are often too expensive, too complex, or require computer network manufacturer hardware and/or software changes. For example, Cloud-Flare's anti-DDoS Enterprise solution starts at 5,000 USD/month.

Our research problem context is limited to DDoS L347 network attacks which traverse the Internet and attack on-line Web servers. We focus on protecting on-line services which require authentication (logging in). There is a great deal of general literature, as how to detect and/or mitigate L347 DDoS attacks. However, the research literature is very weak, concerning how to do this, in our specific research context. Also, much of the literature only answers specific practical questions, but there is a lack of literature concerning the related conceptual and applied research questions. To mitigate all of the L347 network attacks, one must also come also up with the algorithm, as how to mitigate all of these, in a very efficient manner. This requires putting together best in class L3, L4, and L7 specific solutions, into a comprehensive high level system algorithm. Our research question is to design a best in class anti-DDoS L347 solution which costs almost nothing, which is simple to implement, easy to understand and does not require any network equipment hardware or software changes.

1.2 Contributions

As related to L347 attacks against the Bank, we answer various conceptual, applied and practical questions in this paper. In security, there is a defense in which the risk is transferred. As a conceptual design principle, we propose that (1) whenever possible, the Bank transfers DDoS risks to service providers at a very low cost, (2) the Bank uses different IP addresses for Web services, one for pre-authentication and one for post-authentication, and (3) the Bank gives each customer their very own unique sub-domain, which is used by the customer after authentication, to access the bank's Web services.

The above design contributions and reasons for them are explained later in this paper. In addition to the previous design contribution, we have other design contributions which are found in our design cycle discussions.

1.3 Research Methodology

We followed the design science research (DSR) methodology [14]. A DSR IT artifact can also be the design guidelines for an IT artifact, as opposed to a physical IT artifact itself. Our high level IT artifact is our proposed design guidelines and algorithms, which greatly mitigate any and all L347 network based DDoS attacks. Via DSR, an IT artifact should be created, then evaluated, and then re-designed with improvements (based on the feedback from the evaluation). This cycle is then repeated several times. These cycles then continue, until an adequate level of new knowledge is acquired and/or a practical solution emerges.

It turns out that this approach will make it easier for the reader to understand our final and total solution.

1.4 Summary of Network DDoS Attacks

We will provide a very brief overview, of DDoS attacks. Direct network attacks and indirection Reflection Attacks are shown in Fig. 1.

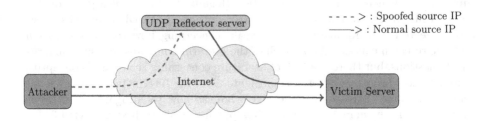

Fig. 1. Direct and reflection attack

An example of how IP source address spoofing works, is shown in Fig. 2.

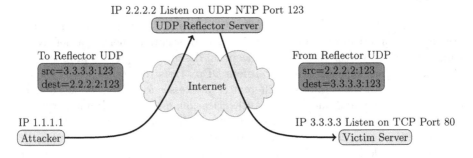

Fig. 2. Packet transitions during UDP reflection attack

With a DDoS attack, there is the attacker controller, masters, Zombies, Reflectors and Victims, as shown in Fig. 3.

1.5 Outline of This Paper

The rest of this paper is organized as follows: In Sect. 2, we perform DSR methodology and go through several design cycles. This is where we explain the specific research problem issues and our proposed solutions. In Sect. 3, we analyze related works and include a synthesis of those works. In Sect. 4, we provide our conclusions and future work suggestions.

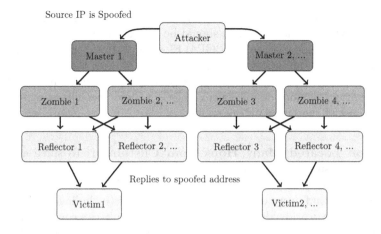

Source IP is Spoofed

Fig. 3. Detailed and complex reflection attack

2 DSR Methodology - Design Cycles

We now present our DSR design cycles and related design contributions.

1. On Premise Solution: We will first provide our design based on a very simple case. Assume that the bank has an on premise only solution, that the bank's ISP link is 10 Gbps, and that there is an attack of 15 Gbps. The only way to defend against this is for the bank to increase their ISP like to, for example 20 Gbps, which may take several days. However, the attacker will perhaps just immediately increase the attack to 30 Gbps. So the bank can not provide a defense, which is entirely on premise.

2. Local ISP Solution: The bank decides to try and mitigate the attack, at their ISP. Some ISPs scrub out malicious DDoS traffic in-house. However, most ISPs do not offer an in-house solution. So most ISPs outsource the scrubbing dynamically after detection, by rerouting customer traffic, via BGP4. The outsource solution might take an hour, after the attack is detected, to start diverting all traffic (via BGP4), to the outsource scrubbing service. The attackers could repeated stop the attack after the outsource is operational. Then they could wait for the traffic to be sent directly to the customer at which time they start the attack again. Both types suffer from an inability to detect all malicious and all valid traffic. So most of the time, some malicious traffic is let through and some valid customer traffic is dropped.

3. Upstream of ISP Solution: In the previous cycle, the bank had the ISP try to filter the attack, before the malicious traffic reached the bank. The ISP can use a similar strategy. The ISP can try to have it's upstream neighbors filter the malicious traffic, before it reaches the ISP. There is a standardized way for the ISP to state which malicious traffic should be filtered upstream, which is called border gateway protocol version 4 (BGP4) Flow-Spec. However, most ISPs don't support BGP4 Flow-Spec. Let's assume the bank is sharing their ISP connection for (1) incoming Web service traffic and (2) outgoing general employee traffic. In this scenario, it is not possible to create BGP4 Flow-Spec filters, which would filter out most of the DDoS attack traffic.

4. Solution Located Many Hops from Customer: To solve all of the previous cycle issues, let's first create a theoretical and virtual solution. We now assume that there is one really great low cost virtual ISP in the world, that can be used by each and every end customer in the world (including the bank). We assume that every worldwide customer, such as the bank, can connect directly to this virtual ISP. So the bank can simply connect directly to this virtual ISP via a 10 Gbps and the virtual ISP can filter out most of the malicious traffic, at a low cost. We have found an actual current solution, which has many similar features, to this virtual ISP solution. However, instead of this virtual ISP connecting to the bank in-line, we need to use a variety of different technologies and completely change the design.

To understand the rest of this paper, one must have a good understanding of how content delivery networks (CDN) and reverse Web proxies work, a great tutorial which is found here [16]. As an example, to illustrate our solution, we will discuss the CloudFlare CDN solution, which has 86 data centers. To use this CDN, in order to gain the above virtual ISP functionality, a summary of the changes follow (which is simplified).

Let's assume that the bank's Web DNS entry currently points to the bank's public and well known IP address, of 11.1.1.1. The bank's Web DNS entry IP address needs to be changed to CloudFlare's IP address of 22.2.2.2. The bank needs to change their previously public IP address of 111.1.1.1 to some secret public IP address, such as 33.3.3.3. This CloudFlare service will setup a reverse proxy in each of their 86 data centers. The bank's customers and attackers will now receive the CloudFlare IP address of 22.2.2.2, when trying to connect to the http//bank.com DNS name. It is important to understand that all 84 CloudFlare data centers will have the same IP address of 22.2.2.2, which is considered anycast routing.

When the bank's customers try to connect to bank.com/22.2.2.2, they will be sent to the closest CloudFlare data center's reverse proxy. Then the CloudFlare reverse proxy will connect to the bank's secret IP address, which is 33.3.3.3. Customers and attackers will never know the bank's secret IP address, so they will never communicate directly with this address of 33.3.3.3. Let's assume that the attackers are sending a 400 Gbps DDoS attack to the bank. Again, via the DNS name of bank.com, they will only learn the CloudFlare IP address of 22.2.2.2. The attackers may use hundreds or thousands of zombies, which are part of a BotNet. So these zombies may be scattered around the world. For each zombie, their attack traffic will be directed to the CloudFlare data center, which is located the closest to this zombie. In summary, the 400 Gbps attack traffic is distributed among the 86 CloudFlare data centers, so the attack at each CloudFlare data center is normally much smaller than the aggregate 400 Gbps.

The majority of high volume bandwidth DDoS attacks are of the type, reflection attacks, which we fully described in our previous International Conference on Future Network Systems and Security (FNSS) conference contribution [4] and a related journal article [5]. This paper picks up where those papers left off. A reflection summary is simply that the attacker can send an aggregate of, for example 10 Gbps, and the reflectors will amplify that by, for example 40x. So the reflectors will receive 10 Gbps from the attackers but they will send 400 Gbps to the victim servers. This 400 Gbps of reflection attack is not valid HTTP traffic, so no possible reflection attack traffic would be forwarded to the bank's secret IP address. So none of the DDoS reflection attack traffic, with a bandwidth of 400 Gbps would ever reach the bank's local premise!

The bank, other organizations and individuals can get this functionality from CloudFlare for free. Now CloudFlare would not be very happy to keep handling 400 Gbps bandwidth attacks without any service revenue. However, CloudFlare can (and does) use BGP4 Flow-Spec. This can be used by CloudFlare, to require their upstream neighbors to filter all non-Web traffic, which is destined to reach CloudFlare's Web reverse proxy servers. So CloudFlare does not need to receive any of this reflection DDoS attack 400 Gbps traffic. Having said that, CloudFlare can instead accept the DDoS attack for a short while, before sending the BGP4 Flow-Spec request, which will allow CloudFlare to better analyze the attack, and to properly rate the agreement Gbps volume.

Of course, the attackers can try other ways to send very high bandwidth DDoS attacks, its just that the reflection DDoS attacks will not be received by the bank's Web server (33.3.3.3). From the perspective of the bank and Cloud-Flare, all reflection attacks have been eliminated. However, in security terminology, we say that this attack risk has been transferred to CloudFlare's upstream neighbors, who will perform the filtering, based on receiving BGP4 Flow-Spec filter requests.

5. Solution for L3 Non-reflection Attacks: L3 network attacks can be either reflection attacks, or non-reflection attacks. In the previous cycle, we basically eliminated DDoS reflection L3 network attacks (from the point of view from CloudFlare and the bank). Without these attacks, we only have L3 non-reflection attacks left. L3 non-reflection attacks can be divided into two types, with IP source address spoofing and without spoofing. The only type of attack traffic that would reach the bank's Web site (33.3.3.3) are valid L4 and/or L7 traffic (which we do not consider as also an L3 attack). So for either type of L3 attack, none of this L3 network attack traffic would reach the bank's Web site (33.3.3.3). All of this L3 attack traffic, would be terminated at or before the CloudFlare reverse proxy. So L3 attacks have no direct effect, on the bank's Web servers. This L3 attack risk has been transferred from the bank to CloudFlare.

6. Solution for L4/7 Attacks: The only remaining attack traffic to consider are L47 attacks. We'll now consider L4 attacks, which we will consider as follows. L4 attacks are based on L4 TCP connection/termination requests, setting TCP flags, and performing strange or unusual TCP transport activities, often at a very rapid rate. For any attack that actually sends traffic to the Web server process, we will consider that later in this paper, as an L7 attack. There are of course attacks, which are both an L4 and L7 attack. We'll now address pure L4 attacks and the L4 portion of any L4+7 attacks. For the attacks that don't end up opening an L4 TCP transport connection, the CloudFlare reverse proxies will not open a TCP connection to the bank's Web server.

Many of the other L4 attacks can be stopped with a standard stateful firewall. CloudFlare's professional plan, which is 20 USD/month, includes a Web application firewall (WAF), which includes support to stop most L4 attacks. An alternate solution is the following, where the bank continues to use the Cloud-Flare free plan. Then, instead of having the bank Web server on premises, at 33.3.3.3, the bank runs their own virtual machine (VM) guest Web server, in the Microsoft Azure cloud. Then CloudFlare is the front-end, for this Microsoft Azure cloud based bank Web server. The low end Azure cloud cost is about 20 Euro/month for a VM guest. This Azure service includes a free L4 stateful firewall, which will stop most of the L4 attacks. Also, the VM guests include a 10 Gbps link, which will handle DDoS L47 bursts in traffic, at a very low cost. The attacks can still spoof their source IP address, to that of a valid session. However, they would need to know the state of the L4 connection, since otherwise the stateful firewalls would block the malicious traffic. If the attack knows the L4 connection state, they can for example, keep sending the most recent TCP response, which could be forwarded to the bank's Web server. However, we'll consider this as an L7 attack, which will be addressed next. So the bank's Web server does not directly receive these L4 attacks. All pure L4 attacks risks has been transferred from the bank to CloudFlare, or in the alternate design to Microsoft.

Solution for Remaining L7 Attacks: We now consider the remaining L7 attacks, which require a TCP connection to be opened with the reverse proxy. To complete the TCP three-way handshake, the client would not be able to use IP source address spoofing. However, once the TCP connection is established, IP source address spoofed traffic can be sent. Here is an example of an L7 attack. Numerous attack clients could collectively open millions of TCP connections and slowly request web pages, in order to deplete the bank's Web server memory, and processing power. This would also be an attack against the network bandwidth. However, since the attack is based on L7 requests, we consider this as an L7 attack, instead of an L3 attack.

We will call the Web clients who are accessible the bank's Web server, but have not yet authentication, as pre-authenticated clients. After they login, we will call those clients as authenticated clients. Many organizations use the same URL for both pre-authenticated and authenticated clients. In this case, as these

L4 attacks are performed, even from pre-authenticated clients, it might have an effect on the authenticated clients. Our design guideline, is to use different Web servers, for pre-authenticated and authenticated clients.

For our example, let's assume that the bank also uses the Microsoft Azure Web hosting service, for handling just the authenticated clients. With this hosting service, it is Microsoft who owns and operates the Web server and the bank only receives Web requests which include the bank's URL. With this design, any L347 attacks towards the Microsoft Web server, which don't include the bank's URLs are handled by Microsoft and have limited effect on the bank's cloud based processes. We recommend that the CloudFlare solution (and bank Web server 33.3.3.3) is now only used for the pre-authenticated traffic. Upon authentication, the specific customer should be sent a Web redirect (or via click URL) to move from the CloudFlare IP address to the Azure Web service.

Dynamic Web Redirection Solution: During normal operation, where there is not an attack, we will have all customers surf to the same URL. However, during a DDoS attack, we will redirect all customers (after authentication) to their own unique URLs. We will now design the architecture, so that during a DDoS attack, we can very easily move almost all of remaining possible Azure related attack traffic, from the bank's Azure Web service process to the Microsoft Web server.

Let's assume that the bank has 1,000 customers. We propose that the bank assigns each of these customers a unique 40-character sub-domain name. Let's suppose account ID number 74 is assigned the DNS sub-domain of "0745X4...BE6". In the Azure web server, you could then create a DNS CNAME entry of `http://0745X4...BE6.Bank2.Com`, which points to this Azure site. If the customer tries to access their account information, the URL might be something like: `http://0745X4...BE6.Bank2.Com/account-info`, instead of http:// Bank2.Com/account-info (which would only be used when there is no DDoS attack). The bank should configure the DNS server to prevent any unauthorized zone transfers, since we need to keep these customer sub-domain names secret. The bank then configures the Azure Web server to accept traffic for these 1,000 sub-domains, but not to accept any other Web requests. When a customer authenticates, via the CloudFlare service, they are redirected to the Azure Web server, with their very own secret sub-domain name.

Now let's talk about how to detect DDoS attacks which reach the Bank's Azure Web server process. We create a list for each specific customer ID/sub-domain. If there are 1,000 active customers, we have 1,000 active lists. For a given customer sub-domain list, we keep track of all source IP addresses, that are actively sending traffic to this customer's sub-domain. A customer would normally not login from more than a couple of IP addresses simultaneously. However, a DDoS attack, by definition, would be when a large number of attackers, would be sending traffic. So if, for example, there is incoming traffic from more than ten source IP addresses, to the same customer sub-domain, we have detected an L7 attack. Put another way, we analyze all traffic, to a given customer sub-domain, and looking at that traffic only, we try to figure out if there is a DDoS attack. It is perhaps 1,000 times easier to detect a DDoS attack, since we only consider traffic towards each customer sub-domain, on its own, and then decide if it looks like a normal bank customer's traffic pattern.

Once an attack is detected, for a specific customer, we have a variety of options. Here is one option. As long as possible, do the following (and only until there is a huge amount of attack traffic). Continue to service requests, to this customer ID sub-domain. However, add a random 1–3 s delay per request, before serving the web pages. Hopefully, the attack would come from a large number of IP addresses, which can be retained. It is public information, as to which ISP/AS owns every public IP address.

When the bank decides to stop an attack, they can (1) terminate all of only this customer's sessions, (2) delete this customer's sub-domain, (3) assign the customer a new sub-domain, and (4) register this new sub-domain on the Azure service. Only after a new successful login, would the customer be redirected to their new customer ID sub-domain, on the Azure service. By deleting the old domain, the following will occur. For all future attack traffic, to this customer's old sub-domain, it would no longer reach the bank's Azure Web process. Instead, it would be the Microsoft Web server, which would be forced to process and filter/drop this attack traffic. So we have also transferred this risk/issue from the bank to Microsoft.

With the above in place, we can now optimize our solution. Microsoft will charge by the minute, for each of the 1,000 Web sites. For 1,000 Web sites, it costs 1,000 times as it would cost for one Web site. Most banks would only be under attack, less than 1 % of the time. So we recommend that when the bank is not under attack, they have just one Azure Web site. When the attack is active, they can have 1,000 Web sites. If the bank wants to save money, where there is an attack, they can instead of ten groups of 100 customers, on ten Web sites. Or they can have one hundred groups of ten customers. For the groups that have an attack, they can then divide the group into ten new groups and redirect the customers to these new groups. For the groups that don't have any attack, they can put these groups back together, in bigger groups.

3   Related Work and Synthesis

We will first present a few comments, concerning the most relevant works and then provide a synthesis, in a table. For the following papers, any of our comments will begin with "**comments:** ".

In [4], we (Booth/Andersson) found a way to mitigate some UDP DDoS reflection attacks. **Comments:** However, if the attackers directly attacked our TCP ports, for the services we were running on each server, we offered no defense.

In [5], we (Booth/Andersson) extended our solution to stop some UDP and some TCP reflection DDoS attacks. **Comments:** However, again, if the attackers directly attacked our TCP ports or directly attacked our UDP ports, for the services we were running on each server, we offered no defense. This paper you are reading now, has continued building knowledge, I.E. improving the mitigation of all DDoS attacks, where our previous papers left off.

In [6], Chonka et al. present that one of the most serious threats to cloud computing itself comes from HTTP Denial of Service or XML-Based Denial of Service attacks. They present their Cloud TraceBack (CTB) solution to find the source of these attacks. **Comments:** Our traceback solution is so much better, since we know the specific customer sub-domain compromised and we have the list of all the non-spoofed source IP addresses, against this specific customer.

In [7], Chung et al. present a way to detect the vulnerable servers, which are used in the DDoS reflection attacks. **Comments:** Our solution simply transfers all reflection attack risks from the Bank to CloudFlare, at no cost.

In [19], Rai and Selvakumar have some up with an algorithm to detect DDoS attacks using the existing machine learning techniques such as neural classifiers. **Comments:** Their problem is that they are analyzing all incoming DDoS attack traffic, together, in one huge messy context. Our solution is much better, since we created an architecture, so that we can analyze incoming attack traffic, against a given customer, in its own customer context. With our approach, it becomes perhaps 1,000 times easier to identify any DDoS attack. In summary, with our approach, we basically have eliminated the usefulness of any, let's analyze all L347 attack traffic, in the global context approaches.

Here are some more of those, let's analyze all incoming DDoS attack traffic, in one huge messy context: [13, 24, 26, 30].

In [29], Yang and Yang propose a new hybrid IP traceback scheme with efficient packet logging to help locate attack hosts which are spoofing their IP addresses. **Comments:** With our contribution, it becomes extremely simple to perform traceback, concerning any attack traffic which reaches the banks' Azure Web process, since the IP address can't be spoofed. However, their solution is perhaps interesting to CloudFlare, since they must defeat the spoofing DDoS attacks (not the bank).

A variety of surveys are available, to help understand the DDoS research topic, such as [2, 3, 12, 15, 17, 21, 23, 25, 27, 31].

In [11], Furfaro et al., propose a DDoS simulator, which can be used to analyze various proposed anti-DDoS algorithms. **Comments:** This should be very useful to WAF vendors to test different anti-DDoS proposed algorithms, before they are put into production.

In [10], Fachkha et al., proposes to characterize Internet-scale DNS Distributed Reflection Denial of Service (DRDoS) attacks by leveraging the darknet space. They empirically evaluate the proposed approach using 1.44 TB of real darknet data collected from a/13 address space during a recent several month period. Their analysis reveals that the approach was successful in inferring significant DNS amplification DRDoS activities including the recent prominent attack that targeted one of the largest anti-spam organizations. **Comments:** It would be interesting for us to implement our proposed solution in the darknet, in addition to using actual beta customers.

In [9], Dietzel et al., study the use of Internet Exchange Points (IXPs)to black-hole DDoS traffic at upstream providers. They find that the research community has been unaware that IXPs have deployed black-holing as a service for their members. Within a 12-week period they found that traffic to more than 7, 864 distinct IP prefixes were black-holed by 75 ASes. **Comments:** Black-holing will also block all valid traffic. In our solution, we have found a way to greatly mitigate any and all L347 attacks, without any required black-holing of valid traffic.

In [28], Yan et al., explore how to defend against DDoS via recent advances in software-defined networking (SDN). They provide a comprehensive survey of defense mechanisms against DDoS attacks using SDN.

In [20], Santanna et al., study Booters, which are DDoS attack platforms as a service, which can be rented, starting at one USD. As a consequence, any user on the Internet is able to launch attacks at any time. In this paper they extend the existing work by providing an extensive analysis on 15 distinct Booters. **Comments:** This is promising since they have an enormous about of actual attack traffic. Once this paper is accepted, we plan to immediately contact them, so that we can analyze how to design will perform against their collected actual DDoS attack traffic.

We'll now analyze the above and other references, via the following specific criteria:

1. Provides strong background, case study and/or survey about DDoS issues?
2. Anti-DDoS Solution?
3. If DDoS solution, can it utilize upstream assistance?

Table 2. Analysis of research categorized by our research criteria categories

Item	Cite	1	2	3	4	Item	Cite	1	2	3	4
0	This paper	✓	✓	✓	✓	14	[18]	✓	✓	✓	
1	[1]		✓	✓		15	[19]		✓	✓	
2	[2]	✓				16	[20]	✓			
3	[3]			✓		17	[21]	✓			
4	[4]		✓			18	[22]		✓	✓	
5	[5]			✓		19	[23]	✓	✓	✓	
6	[6]	✓	✓	✓		20	[24]	✓	✓		
7	[7]		✓	✓	✓	21	[25]		✓		
8	[9]	✓	✓	✓		22	[26]		✓	✓	
9	[10]	✓				23	[27]	✓	✓	✓	
10	[11]	✓	✓			24	[28]	✓	✓	✓	
11	[12]		✓	✓		25	[29]	✓	✓	✓	
12	[13]		✓	✓		26	[30]	✓			
13	[17]		✓			27	[31]	✓			

4. If DDoS solution, does it attempt to remove just the attack traffic, out of line, from authenticated sessions?

We created Table 2, based on our criteria. The citation column (as always) has click-able links to the bibliography. The first item, item 0, is referring to this contribution.

4   Conclusion and Future Work

We have described the research problem as that there are numerous successful DDoS L347 attacks, and that almost all Information Systems are vulnerable. There is an abundance of academic papers, which can detect one type of DDoS or provide mitigation for one type of DDoS. We were unable to find any academic papers or practical solutions, which described a complete, easy to implement, and low cost solution, for organizations who wish to greatly mitigate any and all L347 DDoS attacks, against Web services.

Our hybrid research contribution design filters most of the general attacks, via the free CloudFlare solution. The Microsoft cloud and Microsoft Web server then filters out all of the remaining general attacks. Then within our cloud Web process, we can very easily detect any DDoS and eliminate the DDoS by deleting the attacked sub-domain. We even know which customer is associated with each and every DDoS attack on the Azure Web process.

Our solution will significantly reduce the false positives, as compared to the major anti-DDoS solutions, which are extremely expensive. We can also create lists of known malicious source IP addresses, and share that information with whoever is interested. Our design is extremely low cost and easy to implement solution, which greatly mitigates all of these L347 threats.

Note that this paper is only a conceptual design and the experiment has been left as recommended future work. As future work, we are planning to implement our solution, put it into production, and publish the related case studies. We are actively searching for volunteers, who wish to participate in our experiments. Other future work is to also come up with other similar solutions, for protocols other than HTTP and HTTPS.

References

1. Alwabel, A., Yu, M., Zhang, Y., Mirkovic, J.: SENSS: observe and control your own traffic in the internet. In: Proceedings of the 2014 ACM Conference on SIGCOMM, SIGCOMM 2014, pp. 349–350. ACM, New York (2014)
2. Arukonda, S., Sinha, S.: The innocent perpetrators: reflectors and reflection attacks. Adv. Comput. Sci. **4**, 94–98 (2015)
3. Bhuyan, M.H., Bhattacharyya, D.K., Kalita, J.K.: An empirical evaluation of information metrics for low-rate and high-rate DDoS attack detection. Pattern Recogn. Lett. **51**, 1–7 (2015)
4. Booth, T.G., Andersson, K.: Elimination of DoS UDP reflection amplification bandwidth attacks, protecting TCP services. In: Doss, R., Piramuthu, S., ZHOU, W. (eds.) FNSS 2015. CCIS, vol. 523, pp. 1–15. Springer, Heidelberg (2015)

5. Booth, T., Andersson, K.: Network security of internet services: eliminate DDoS reflection amplification attacks. J. Internet Serv. Inf. Secur. (JISIS) **5**(3), 58–79 (2015)
6. Chonka, A., Xiang, Y., Zhou, W., Bonti, A.: Cloud security defence to protect cloud computing against HTTP-DoS and XML-DoS attacks. J. Netw. Comput. Appl. **34**(4), 1097–1107 (2011)
7. Chung, C.-J., Khatkar, P., Xing, T., Lee, J., Huang, D.: NICE: network intrusion detection and countermeasure selection in virtual network systems. IEEE Trans. Dependable Secur. Comput. **10**(4), 198–211 (2013)
8. CloudFlare. 400gbps: Winter of Whopping Weekend DDoS Attacks. https://blog.cloudflare.com/a-winter-of-400gbps-weekend-ddos-attacks. Accessed 2 May 2016
9. Dietzel, C., Feldmann, A., King, T.: Blackholing at IXPs: on the effectiveness of DDoS mitigation in the wild. In: Karagiannis, T., et al. (eds.) PAM 2016. LNCS, vol. 9631, pp. 319–332. Springer, Heidelberg (2016). doi:10.1007/978-3-319-30505-9_24
10. Fachkha, C., Bou-Harb, E., Debbabi, M.: Inferring distributed reflection denial of service attacks from darknet. Comput. Commun. **62**, 59–71 (2015)
11. Furfaro, A., Malena, G., Molina, L., Parise, A.: A simulation model for the analysis of DDOS amplification attacks. In: 17th USKSIM-AMSS International Conference on Modelling and Simulation, pp. 267–272 (2015)
12. Gillman, D., Lin, Y., Maggs, B., Sitaraman, R.K.: Protecting websites from attack with secure delivery networks. Computer **48**(4), 26–34 (2015)
13. Giotis, K., Androulidakis, G., Maglaris, V.: A scalable anomaly detection and mitigation architecture for legacy networks via an OpenFlow middlebox. Secur. Commun. Netw. **9**, 1958–1970 (2016)
14. Hevner, A.R., March, S.T., Park, J., Ram, S.: Design science in information systems research. MIS Q. Manag. Inf. Syst. **28**(1), 75–105 (2004)
15. Nexusguard: Whitepapers on DDoS Mitigation, Cyber Attack. https://www.nexusguard.com/genius/whitepapers. Accessed 20 Apr 2016
16. Nygren, E., Sitaraman, R., Sun, J.: The Akamai network: a platform for high-performance internet applications. SIGOPS Oper. Syst. Rev. **44**(3), 2–19 (2010)
17. Osanaiye, O.A.: Short Paper: IP spoofing detection for preventing DDoS attack in Cloud Computing. In: 2015 18th International Conference on Intelligence in Next Generation Networks (ICIN), pp. 139–141, February 2015
18. Poulopoulos, L., Mamalis, M., Polyrakis, A.: FireCircle: GRNET's approach to advanced network security services' management via BGP flow-spec and NET-CONF. In: 2012 Proceedings of the 28th TERENA Networking Conference (2012)
19. Raj, K., Selvakumar, S.: Distributed denial of service attack detection using an ensemble of neural classifier. Comput. Commun. **34**(11), 1328–1341 (2011)
20. Santanna, J.J., Durban, R., Sperotto, A., Pras, A.: Inside booters: An analysis on operational databases. In: 2015 IFIP/IEEE International Symposium on Integrated Network Management (IM), pp. 432–440, May 2015
21. Santanna, J.J., van Rijswijk-Deij, R., Hofstede, R., Sperotto, A., Wierbosch, M., Granville, L.Z., Pras, A., Booters; An analysis of DDoS-as-a-service attacks. In: 2015 IFIP/IEEE International Symposium on Integrated Network Management (IM), pp. 243–251, May 2015
22. van der Steeg, D., Hofstede, R., Sperotto, A., Pras, A.: Real-time DDoS attack detection for Cisco IOS using NetFlow. In: 2015 IFIP/IEEE International Symposium on Integrated Network Management (IM), pp. 972–977, May 2015

23. Steinberger, J., Sperotto, A., Baier, H., Pras, A.: Collaborative attack mitigation and response: a survey. In: 2015 IFIP/IEEE International Symposium on Integrated Network Management (IM), pp. 910–913. IEEE (2015)
24. Thatte, G., Mitra, U., Heidemann, J.: Parametric methods for anomaly detection in aggregate traffic. IEEE/ACM Trans. Netw. **19**(2), 512–525 (2011)
25. Usha Devi, G., Priyan, M.K., Vishnu Balan, E., Gokul Nath, C., Chandrasekhar, M.: Detection of DDoS attack using optimized hop count filtering technique. Indian J. Sci. Technol. itextbf8(26) (2015)
26. Xiang, Y., Li, K., Zhou, W.: Low-rate DDoS attacks detection and traceback by using new information metrics. IEEE Trans. Inf. Forensics Secur. **6**(2), 426–437 (2011)
27. Yan, Q., Yu, F.R.: Distributed denial of service attacks in software-defined networking with cloud computing. IEEE Commun. Mag. **53**(4), 52–59 (2015)
28. Yan, Q., Yu, F.R., Gong, Q., Li, J.: Software-defined networking (SDN) and distributed denial of service (DDoS) attacks in cloud computing environments: a survey, some research issues, and challenges. IEEE Commun. Surv. Tutor. **18**(1), 602–622 (2016)
29. Yang, M.-H., Yang, M.-C.: RIHT: a novel hybrid IP traceback scheme. IEEE Trans. Inf. Forensics Secur. **7**(2), 789–797 (2012)
30. Yu, S., Zhou, W., Jia, W., Guo, S., Xiang, Y., Tang, F.: Discriminating DDoS attacks from flash crowds using flow correlation coefficient. IEEE Trans. Parallel Distrib. Syst. **23**(6), 1073–1080 (2012)
31. Zargar, S.T., Joshi, J., Tipper, D.: A survey of defense mechanisms against distributed denial of service (DDoS) flooding attacks. IEEE Commun. Surv. Tutor. **15**(4), 2046–2069 (2013)

Secure RFID Protocol to Manage and Prevent Tag Counterfeiting with Matryoshka Concept

Gaith Al.[1]([✉]), Robin Doss[1], Morshed Chowdhury[1], and Biplob Ray[2]

[1] School of Information Technology, Deakin University, Geelong, Australia
galiyev@deakin.edu.au
[2] School of Engineering Technology, CQ University, Rockhampton, Australia

Abstract. Since the RFID technology has been found couple of decades ago, there was much involvement of this emerging technology in the improvement of supply chain management. As this technology made the industry more reliable and faster to process, yet there were always some technical issues and security threats that emerged from the heavy use of the RFID tags in the SCM, or other industries. Hereby we represent a new protocol based on a new idea that can be used to manage and organize tags as well as the objects attached to them in SCM, to prevent counterfeiting and reduce the security threats taking into consideration the security and privacy concerns that faces the industry today. This new approach will open a new horizon to the supply chain management as well as the RFID systems technology since it will handle multi- tags attached to objects managed in one location as an entity of one in one. We called our approach the MATRYOSHKA approach since it has the same idea of the russian doll, in managing multi-tags as one entity and prevent counterfeiting. We also added extra authentication process based on a mathematical exchange key formation to increase the security during communication to prevent threats and attacks and to provide a secure mutual authentication method.

Keywords: RFID · Matryoshka · SCM · Anti-counterfeiting

1 Introduction

The RFID system is a system which uses radio frequency (RF) to communicate between receivers and transponders. Every RFID system has three basic components, tags (transponders), readers (receivers) and data base. The tags usually are attached to objects, human, animals, etc. to be identified through Information stored in the database. There are three types of tags, active, semi-active and passive. The passive tags are the most commonly used since it does not need an internal power source and it is very low in cost and easy to manufacture [1]. On the other hand the supply chain management (SCM) is the "management and control of all goods and information in the logistics process from acquisition of raw materials to delivery to the customer" [2].

© Springer International Publishing AG 2016
R. Doss et al. (Eds.): FNSS 2016, CCIS 670, pp. 126–141, 2016.
DOI: 10.1007/978-3-319-48021-3_9

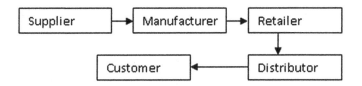

Fig. 1. The supply chain

The Supply chain covers all processes involved in the flow of stocks and goods from factory to customer; including manufacturing, storing, distribution and transportation. The use of RFID in supply chain can be considered an important and significant step forward in managing and controlling the flow of the good and materials from the manufacturer to the customer after going through all other steps between. As shown in Fig. 1.

RFID technology implemented in the supply chain was proven effective and much more reliable than its previous older version brother the barcodes since it ensures the goods are available in the right place with no discrepancies and zero errors [3].

The technology has increased the accuracy, reliability and efficiency of the supply chain and made it more effective by providing instant information's for the goods attached to the tags, which make the supply chain process much easier to manage, control and secure. In general we can summarize the benefits of using the RFID technology in SCM with the following points.

- Increase accuracy
- Low cost.
- Reduce the manual work.
- Provide real time information for the product.
- Faster product location detection.
- Improve the process speed.
- improves efficiency.
- Improve the planning of the stock flow.
- Improve visibility for the products.
- Improve tractability for the products.
- Improve security.

Yet there is also some disadvantages which occurs from time to time while using the RFID systems in SCM, we state some of them below:

- Interference and reading possible issues
- Software and equipments upgrade and maintenance
- Lack of slandered
- Privacy concerns

Frequently especially during transportation of good and packages, there is always a risk of losing or damaging tags during transportation of the stock even if they were physically packet together in a proper way from the source. This random issue will lead to confusion in reading the stock not mentioning the huge amount of disruption which can be caused by thousands or millions of tags, packeted together in one container during the reads or during the attacks which might occur during the transportation of the goods or packets in the SC hubs.

We believe that in our proposed protocol "Matryoshka" we can minimize these problems and issues to the minimum by establishing a system which adopt a mechanism that can group thousands or millions of tags to be presented by few tags or one tag instead. Which will lead us to minimize the reads to the minimum to avoid the problems mentioned above. In the next section we will detail some of the disadvantages which might occur when using the RFID in SCM. As it's important to point out some of these issues to understand the benefits and the strong aspects of our new protocol which has a multi tag levels to organise the stocks and the tags in a way that benefits the SCM. RFID protocols usually will prefer less complexity in tag computation, less communication between tags and readers, to deal with the scalability issue and to improve the look-up process in the tag table [2].

In our opinion the strong aspects of this paper are as follow: providing a new mechanism in grouping RFID tags in SCM taking security, anti-counterfeiting and privacy into account, reducing the communication between the tags and readers by presenting one tag instead of many which will lead to reduce the scalability issue of RFID tags to the minimum and applying a new secret key exchange method between tags and readers that is stronger and more reliable mathematically than the previous methods as we will see later. While the weak aspect of this method in our opinion will be the increase risk of physical tag removal and suspicious human intervention that might cause incorrect performance of our proposed protocol. We named our protocol Matryoshka in managing multi-tags after the russian doll Matryoshka, also known as the Russian nesting doll or babushka doll, refers to a set of wooden Dolls of decreasing size placed one inside the other [4]. Since the idea is similar in some ways, we thought it will be easier and more helpful to use the Matryoshka name to refer to our new approach in managing multi tags since it's easier to remember. Brad Hodson once wrote, "Just like a Matryoshka doll, each part of your process should fit neatly into another part of the process, and so on until everything is safe and sound in one container. But not just together like that old box of LEGOs you used to play with. Each piece needs to neatly fit into the next, in the right and same order every single time, without mistakes. Just like a russian doll".

2   Literature Review in the Use of RFID System in the Supply Chain Management (SCM)

The use of RFID systems in supply chains management (SCM) used to be and always been a revolutionary method that increases the efficiency and effectiveness of the SCM. In a study conducted by [5] the researchers addresses the strategic values and challenges that effect the efficiency of RFID technology in

SCM. They mentioned that the major goals of deploying RFID in SCM are: authentication, location, and automatic data acquisition (ADA). Also they had concluded that companies when they develop their RFID strategies must look beyond more compliance for ways to implement initiatives into their total supply chain strategy, to harness the true business value of the technology and increasing profits. Never the less we have to take into our consideration when dealing with SCM process the eight key processes and their functions as mentioned by [6,7], Those eight key processes are:

1. Customer relationship management.
2. Customer Service Management.
3. Demand Management.
4. Order Fulfillment Management.
5. Manufacturing Flow Management.
6. Supplier Relationship Management.
7. Product Development and Commercialization.
8. Returns Management

Since number five above "Manufacturing Flow management" requires the managing of products flow, we can say that our protocol will target this point in future work. Michael and McCathie [8] highlighted the Pros and cons in managing multi-tags in RFID supply chain. She elaborated on the pros of the RFID system in SCM, mentioning some of the features that the technology provides such as: Non-Line-of-Sight Technology, Labor Reduction, Enhanced Visibility and Asset Tracking and Returnable Items. Also she highlighted some of the cons while using the RFID system in SCM such as: Privacy Concerns, Lack of Standards, Deployment Issues, Interference and Reading Considerations. On the last point Katina mentioned the following "since RFID uses the radio spectrum to transmit its signals, it is susceptible to interference, hindering its ability to transmit clear and reliable information to RFID readers Similarly, RFID suffers from the inherent range limitations associated with the radio spectrum" [8]. McGinity stated in his paper that RFID system has the ability to read through most packaging materials such as plastic wraps and cardboard containers is one of its most valuable assets. Metal and liquid have been described as the "kryptonite to RFID" as they can play havoc with RFID signals. Ari Juels, presented a similar idea to our Matryoshka protocol. His Idea was based on a Yoking- Proofs for RFID Tags [9]. The Aim of this Idea was to enable a pair of RFID tags to generate a proof that is readable simultaneously by a reader. The author referred to this as a yoking-proof meaning to join together and he also present a basic protocol for tags capable of basic cryptographic operations like MAC and keyed hash functions. The paper [9] gave a brief examples on where this is useful, mentioning legal requirement in pharmaceutical distribution were one RFID tag is embed in a container for the medication while another is embedded in an accompanying leaflet. Later Leonid Bolotnyy and Gabriel Robins In their paper "Generalized Yoking- Proofs for a Group of RFID Tags" [10], presented a protocol that creates a proof for large group of RFID tags which are read within specified time bound and assumed a trusted offline verifier which does not need to verify the proof directly once its created neither communicate with tags. Yet the main Idea of yoking Proof is unlike our presented Matryoshka protocol dealing with less number of RFID tags while our idea was established to include and group a large number of tags together in huge galaxies like clusters as we will see later. There were also many security protocols that address scalability which was classified by [11] into: delegation technique such as [12], tree based approach such as [13], group- based approach such as [14] and collaborative approach such as [15]. The approach which proposed by [11] show that there is a tremendous increase in computational complexity when the numbers of tags increases. Those scalable protocols can be replaced by Matryoshka which will decreases the tags communication to the minimum.

3   Problems in Using RFID Systems in (SCM)

Since the RFID use in SCM has many advantages, there are also many problems which occurs, such problems can be very simple but also catastrophic when they were used in areas with high level of importance. The RFID industry is increasingly working on developing technologies to handle these problems and make the use of RFID systems in SCM more reliable, affordable and error free. Usually when using an RFID technology in SCM the reader must read every tag in every step to assure the quantity which has received is the same on the other end as well as the information's which is stored in the tags are correct. So many reads required from a huge number of tags in short period of time which might cause one of the following problems:

- RFID systems can be easily disrupted: which mean the RFID systems are easy to jam since it depends on RF signals to communicate.
- RFID Reader Collision: happens when two or more readers signals over laps. This problem causes the tag to fail to respond in the same time, many systems uses anti collusion protocols to avoid this problem.
- RFID Tag Collision: this problem occurs when using many tags in a small area.

Our proposed protocol will reduce the number of the read to the minimum which will leads to reduce the problems mentioned above that will lead in its turn to produce high accurate reading. In the next sections we will discuss our proposed protocol as well as the ways that our protocol can contribute to the reduction of the number of tags reads and to the reduction of the collisions occurred during the reading process as mentioned above. Yet our protocol will have one important condition to operate at all times that is the assurance of no physical disruption during transportation of good or pallets, other wise a confusion might happen and the system might stop working which make this the major weak aspect of the protocol.

4 Protocol Setup

4.1 RFID Tags Levels

To setup the Matryoshka protocol we first have to classify the tags into several levels depending on the stock quantity. Having said that we can organize and define the RFID tags level as:

1. 1- Level 1: In this level all the tags must be attached to the items directly as per the scenario at Sect. 6, also the RFID tags in this level cannot be master tags what so ever.
2. Level 2: The tags in this level supposed to be master tags but they can also act as slave tags in the same time. The best given example in the scenario below where the tags attached to the pallets which holds the Items that are attached to the tags in level 1.
3. 3- Level 3: The tags in this level can be act as master tags only, the best given example in the scenario at Sect. 6 is the tags attached to the trucks or the containers.

4.2 Tags Mute/Un-Mute

We assume that each tag must have a flag which is set to 0 or 1 when the value of a tag is 1 that means that the tag in muted or the reader will discard the tag read but when this flag value is 0 the reader will read the signal generated from that tag. To clarify the logical mute function more we can also classify this function based on the level which use it in our protocol:

1. 1- Logical Mute function: All the tags in level one and level 2 or in other words all the tags which acts as slave tags in certain period of time during the process, will have a mute function. This function orders the tags not to response to any signal until it un-muted.
2. 2- Logical Un-mute function: This function is opposite to the mute function, it can be issued only from the master tag to the slave tags via the trusted reader. It will allow the tags in level 1 or 2 to respond to the readers individually just as they normally do.

4.3 Pyramid Structure

All the tags in this protocol will be placed in a pyramid structure which will allow the system to identify which tag is in level 1, 2 or 3. This pyramid structure will start from the bottom with level 1 slave tags "LS", then in the middle there will be less numbered Level 2 tags also called slave and master tags or "LSM" tags, and in the top of this pyramid there will be the level 3 tags of master tags "LM". This Pyramid like structure will provide a clear idea for the location of each tag in the Matryoshka protocol in order to assist organizing and managing the tags in the protocol based on their levels. See Fig. 2.

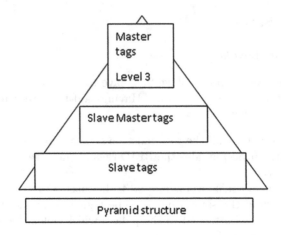

Fig. 2. The pyramid structure for the protocol setup

4.4   No Physical Disruption for the Tags Structure Package

In our protocol there should be no physical disruption at all times, which means that the tag structure should always be secure in a sealed container or boxes. This will assure the tag numbers will be more accurate in the master tag even n the case of the tags was corrupted or damaged. Also it will provide better security to the tags in level 1 since they are all muted. While the tags in level 2 and 3 will have a very complicated tag ID's determined from the calculations in formula 1 and 2 as we will see in the next section.

5   The Matryoshka Protocol

Level 1 tags: The tags ID's in level 1, will be named as $LS(TID)$. The database then will generate a random numeric values for each $LS(TID)$ we will call it $LS(TID)'$ and store this value in a table called LS table (see Table 1 below).

Table 1. The LS table

TID	Random generated number(K)	LS(TID)'
$LS(TID)_i$	k_i	k_i
$LS(TID)_{i+1}$	k_{i+1}	k_{i+1}
$LS(TID)_{i+2}$	k_{i+2}	k_{i+2}
$LS(TID)_{i+n}$	k_{i+n}	k_{i+n}

Table 2. The LSM table

TID	$LSM(TID)'$
$LSM(TID)_i$	$LSM(TID)'_i$
$LSM(TID)_{i+1}$	$LSM(TID)'_{i+1}$
$LSM(TID)_{i+2}$	$LSM(TID)'_{i+2}$
$LSM(TID)_{i+n}$	$LSM(TID)'_{i+n}$

The DB then will determine the tag ID (TID) for each Level 2 tags "$LSM(TID)$" by adding the values of $LS(TID)'_{i+1}$ assigned in LS tables together,

$$LSM(TID)' = \sum_{b=1}^{n} LS(TID)'_b \qquad (1)$$

the readers will assign a new tag IDs for level 2 tags as well based on formula (1) and the tag location in the Matryoshka structure. Then all the tags for Level 1 will be muted. See Figs. 5 and 6. Level 2 tags: At this level all tag IDs will be allocated in a table in the database named LSM table. The Reader will Write the original $LSM(TID)$ values in column 1 and use formula 1 to generate new values for $LSM(TID)'$ and then write those values at column 2 from the LSM table as shown in Table 2 below:

The generated numeric $LSM(TID)'$ value will replace the $LSM(TID)$ in the data base. The data base will determine the $LM(TID)$ of Level 3 by Xoring the values of $LSM(TID)'_{i+1}$ as shown in formula 2 were n is the number of tags, while $i = 1$.

$$LM(TID)' = LSM(TID)'_i \oplus LSM(TID)'_i \oplus n \qquad (2)$$

The reader then will replace the actual $LM(TID)$ with the value of the master tag $LM(TID)$ which supposed to be on top of the pyramid structure. Then all the tags for Level 2 will be muted.

Level 3 tags: In this level the tags will also be named master tags since they are located on the top of the pyramid structure of the Matryoshka protocol. The tags here will represent all the tags located underneath it which will allow the reader to communicate with one tag instead of many since the other tags in level 1 and 2 are muted.

5.1   Retrieving Tags Original ID's Input and Output for Level 1 or Level 2 Tags

Since there is no physical interruption with the packages that include the tags, there must be some way to retrieve the tags into their original values in order to deal with them individually. This will help the stock flow to de-group the tags again at any time. The main element in this procedure is to follow the value of the master tags and allocate them in the LSM or LS tables in order to determine which master tag is allocated in which table.

5.2 Algorithms

For security reasons' we will assume that all tag IDs are hashed at the beginning. The reader (R) will read all the tags in LS level then read the $LSM(TID)_i$ and send the info's to the DB.

Algorithm 1. Begin Algorithm 1
___
1: Read $LS(TID)_i$
2: Send "Mute" to $LS(TID)_i$
3: Read $LSM(TID)_i$
4: Send "Mute" to $LSM(TID)_i$
5: Read $LM(TID)$
6: Call DB1
7: Write $LM(TID)'$ to $LM(TID)$

Algorithm 2. Begin DB1
___
1: Create LS table
2: Write $LS(TID)_i$ to LS table Column 1
3: Generate Random number (K) for LS table
4: Write $(K)_i$ to Column 2
5: $LS(TID)'_i = (K)_i$
6: Write $LS(TID)'_i$ to Column 3
7: Find $LSM(TID)'_i$ from Eq. 1
8: Repeat step 1 n times
9: Create LSM table
10: Write $LSM(TID)_i$ to LSM table column
11: Write $LSM(TID)_i$ to LSM table column 2
12: Determine $LM(TID)'$ from Eq. 2

To retrieve the tags ID's to their original values

Algorithm 3. Begin Algorithm 3 reader
___
1: Read LM(TID)
2: Call DB2
3: Un-Mute LS(TID)i
4: Un-Mute LSM(TID)i

5.3 Tag Authentication Process

In order to authenticate each other's the Master tag and the reader can use a mathematical exchange key formation that was proposed by Stickel [16] as follow: Let G be a non-abelian finite group, a, b belongs to G such That ab not equal ba. Let $n1$ be the order of the element a and $n2$ be the order of the element b.

Algorithm 4. Begin DB2

1: **if** $LM(TID) = (LM(TID)')$ **then**
2:     Determine $LSM(TID)'_i$ from Eq. 2
3:     Determine $LSM(TID)_i$ From LSM table
4:     **if** LS table $N = LM(TID) = (LM(TID)'_i)$ **then**
5:         Determine $LS(TID)'_i$
6:         Determine $LS(TID)_i$
7:     **else**
8:         wrong table
9:     **end if**
10: **else**
11:     wrong value
12: **end if**

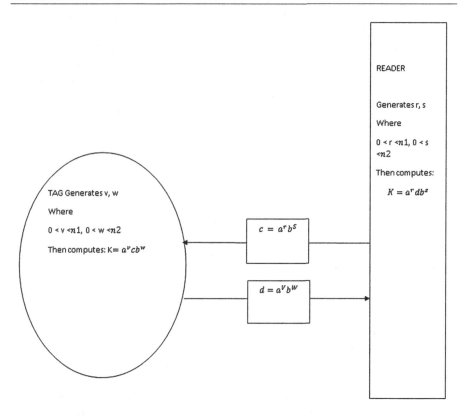

Fig. 3. Mutual authentication process

1. The reader randomly generates natural numbers r and s with $0 < r < n1$, $0 < s < n2$. r and s are kept Secret. Then the reader forms $c = a^\wedge r b^\wedge S$ and sends c to the Tag.
2. The tag after randomly generates natural numbers v and w with $0 < v < n1$, $0 < w < n2$. v and w are Kept secret. Then it forms $d = a^\wedge V b^\wedge W$ and then q sends it back to the reader.

3. The tag computes $K = a^\wedge v\, cb^\wedge w$. k is the secret key used in the subsequent communication between the Tag and the reader.
4. The reader also computes $K = a^\wedge r\, db^\wedge s$.

Both parties will have K known as a secret key which they might use for authentication. See Fig. 3.

6   Scenarios to Understand the Matryoshka Protocol

In order to understand the new idea we would suggest the following scenario; let's imagine a shoe factory which produces sport shoes for some retailers around the country. Let's assume that this factory did receive an order from a retailer to supply him with 300 pairs of shoes in 3 different colors black, white and red, 100 black pairs, 100 white color and 100 red color pairs of shoes. The first 100 pairs are packed in pallet A, the second 100 are packed in pallet B and the third in pallet C. The three pallets then are packet inside container 1 all together. The supplier did attach a tag to each pairs of shoes before moving them to the store; also he did attach a tag to each pallet and to the container. See Fig. 4 below. In the use of the current technology there would be a great chance that one of the problems mentioned above in III might occur. Now let's have a look how this scenario would have worked if we used the Matryoshka protocol. First, while still at the factory, the reader will read all the tags normally as it always do and will input all the information's into its database via the readers. Then the system will determine which one of those tags is attached to pallet A that contains 100 pairs and assign this tag as a master tag to the 100 tags attached to the pair of shoes located and packed on that particular pallet. The system will do the same for pallet B and C. To explain it more, pallet A will include all the 100 black pairs, pallet B will include all the 100 white pairs and pallet C will include all the 100 red pairs. So the reader has to make 3 reads instead of 300. Further more the three pallets A, B, C are packed inside container 1 which has also a tag attached to it so the system will assign this tag as a master tag for the other three master tags attached to the pallets and the system now need to make only one read instead of 300 to know what's there inside the container. In other words by using Matryoshka we need to have only one tag reading (the master tag) to determine the numbers of all the items included there since all the tags IDs are merged in the master tag ID. In another scenario we have a Spare parts store which include a stocks of spare parts such as wheels, set of plugs, car front and rear pumpers, filters, brake pads, etc. These entire item are organized in a set of vertical and horizontal matrix like shelves, we will assume also that a passive tags are attached to each spare parts item; these tags are all registered in the system's database which allows the readers to communicate with each tag and identify them. These items are also subject to input and output due to sell an extra purchases which needs the items quantities to be updated regularly as well as the tags attached to them. This might lead to many challenges and difficulties especially if there were a lot of in and out stock traffic every day.

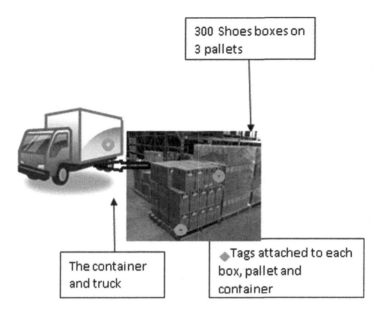

Fig. 4. The tags attached to the boxes on the pallets will be presented by the tags attached to the pallets while the tags attached to the pallets will be presented by the tag attached to the container only

That might cause chaos and difficulty for the system as well as for the workers since they have to update the system every time a change to the stock occurred, the misplaced items can be added to these difficulties since it will be very hard to identify them. While the main issue will remain that the system has to deal with a huge amount of data and tags, as singular items which might lead to the known problems of: (1) RFID system disruption, (2) RFID tag collision and (3) RFID reader Collision Now let's assume that the management wanted to adopt our new method, so they start to attach a passive tag to each shelf which also known to the data base and available to the reader to communicate with. As mentioned above this will also minimize the potential errors and increase the security of the tags since most of them are muted and kept silence, while the master tag will be very well known to the system which increase the security of the RFID system from most of the malicious attacks such as evasdropping, man in the middle, etc.

7   Discussion

Our Matryoshka protocol will provide strong security solutions for many known attacks especially evasdropping and man in the middle attack. Since the level one tags are already muted and presented by level 2 master tag which makes the communication between the tags and readers reduced to the minimum. The reader most of the time will communicate only with the master tag which will

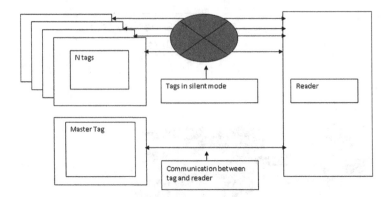

Fig. 5. Communications between master tag and the reader when the Matryoshka been applied

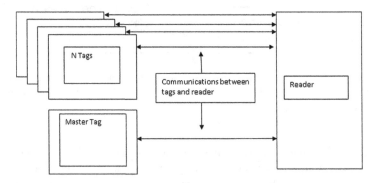

Fig. 6. Communications between tags and reader in normal mode

reduce the chances of RFID system disruptions, tag collision and reader collision. Beside that the Matryoshka can be adapted and developed to produce an extra security protocol to manage the attacks that pause a threat to the system. The idea of reading one tag instead of hundreds, thousands or even millions seems very much promising and revolutionary in our opinion. Yet this Idea as any other new idea need to be developed and tested many times in order to develop a better understanding for the security threat or privacy problem that might occur later. We believe that the Matryoshka protocol will increase the security of the RFID system since the reader will communicate with the master tag only instead of hundred or thousands tags which make it hard for the adversary reader to follow the communication sequence especially when LM(TID)_i is replaced with a value from formula 2. The LS and LSM tables are very much secure in the database since it does not share the contents and values in those tables with a third party, so the values and (TID)' will always be secure. As mentioned above in Sect. 4.4 that this protocol require no physical disruption for the tag structure in the container or the pallet, this would allow the master tag in Matryoshka to provide a very precise value of the numbers of the stock contained in the master tag at the beginning of the shipment or since its last read, even in the case of tag removal, or tag destruction in both level 1 and level 2 in our pyramid structure mentioned in IV section D.

8   Security Analysis

As shown above the protocol will add more security to the tags during the transportation of the goods in the supply chain until it reaches the retailer, by implementing our scheme the protocol will achieve the following security properties:

8.1   Anti-counterfeiting and Cloning

Since all the tags in LS and LSM levels are known only to the DB and since all of these tags won't replay to the readers since they are muted and physically contained. The attacker won't have a chance to counterfeit any of those tags since there was no communication established unless the attacker will know the original (TID) from the DB which is very unlikely to occur. In case the attacker was able to compromise the tags in the LS and LSM the value which are written will not be the true value of the tag that is stored but just another mask value that was obtained from Tables 1 or 2. So even if the attacker will compromise a tag in the transportation process through the SC, the attacker still won't be able to access the tag info's or clone the tag or spoofing it. Since the masked values $LSM(TID)'$ and $LS(TID)'$ will provide extra security for the system and make such kind of attack hard to occur. So the only way to counterfeit the tags will be possible when the master tag is physically removed but still this can be detected at once because there would be no answer from the master tag once the reader interrogates it.

8.2   Tag ID Anonymity

Since the tags are all logically grouped muted during the transportation it won't be possible to track the ID's of the slave tags or Slave Master tags which will provide very strong tag ID anonymity during the whole transportation process, the Master tags ID's $LM(TID)$ can be known as well but it will only be useful for the original Database and the genuine readers which make the possibility of detect or compromising the tag very unlikely. Also the tag will not reveal transmitted data since the communication between the tag and readers will be conducted only in safe environment that have access to the database when retrieving the tags original ID's for level 1, and level 2 which are both $LS(TID)$ and $LSM(TID)$.

8.3   Forward Security

As shown above in Eqs. (1, 2) if the Master tag has been compromised and its current ID has been obtained this will not allow the attacker to trace any previous communication neither determining the true value of the Master tag since it is known only to the DB which it can obtain from Tables 1 and 2, since the value $LM(TID)'$ has been XORed the $LSM(TID)'_i$ as shown in equation.

8.4 Relay Attack

If the attacker try recording and replaying messages from Previous rounds between the Master tag and the reader, the attacker will be unable to establish a communication with the tag as it won't be able to figure out the secret which is used in the authentication process between tags and readers since the used the mathematical exchange key formation that was proposed by [16], this will make replaying messages from the attacker unsuccessful, even if the attacker was able to listen to the communications between master tag and the readers. Despite the fact that it's TID has been changed according to the protocol from Table 2. An attacker cannot impersonate a tag by recording and replaying messages from previous rounds. As the reader issues a fresh challenges for each query so the attacker cannot succeed by replaying an old message as the reader randomly generates natural numbers r and s with $0 < r < n1$, $0 < s < n2$ and the tag randomly generates natural numbers v and w with $0 < v < n1$, $0 < w < n2$ the rest of the tags which are in the pallets will not respond to the integration so this method is only valid for the Master tag and the attacker won't be able to obtain the shared secret k from the master tag since the other variables are keep changing as mentioned in above.

8.5 DoS Attacks

Since the Master tag will be the only tag to replay to the readers during the stock flow in the supply chain this will minimize the DoS attack to the minimum again and make it hard for the attacker to overwhelm the tags with many messages as the Master tag will ignore them all and respond only to the messages from the reader with the exchange key.

9 Conclusion

We presented a new secure method in scalability and managing tags in (SCM) which will provide more accuracy and more reliability in tags security and management. This method will decrees the problems, threats and errors associated with tags reading in RFID systems such as disruption, tag collision, tag counterfeiting threats etc. The reduction of the tag readings can also be very important for the privacy and security and can be adapted for some tag ownership transfer protocols such as [4,5]. We believe that Matryoshka approach will add much to the security of the RFID industry that can be improved and investigated further more to take better shape in the future.

References

1. Al, T., Al, G.K.D.: A case study in developing the ICT skills for a group of mixed abilities and mixed aged learners at ITEP in Dubai-UAE and possible future RFID implementations. In: Luaran, J.E., Sardi, J., Aziz, A., Alias, N.A. (eds.) Envisioning the Future of Online Learning, pp. 133–146. Springer, Heidelberg (2016)
2. Chen, X., Cao, T., Guo, Y.: A new scalable RFID delegation protocol. Appl. Math. **8**(4), 1917–1924 (2014)
3. RFID ARENA. Benefits-of-implementing-rfid-in-supply-chain-management (2015). http://www.rdarena.com/2013/11/14/benets-of-implementing-rd-in-supply-chain-management.aspx
4. AL, G., Ray, B., Chowdhury, M.: RFID tag ownership transfer protocol for a closed loop system. In: 2014 IIAI 3rd International Conference on Advanced Applied Informatics (IIAIAAI), pp. 575–579, August 2014
5. Al, G., Ray, B., Chowdhury, M.: Multiple scenarios for a tag ownership transfer protocol for a closed loop system. IJNDC **3**(2), 128–136 (2015)
6. Cooper, M.C., Lambert, D.M., Pagh, J.D.: Supply chain management: more than a new name for logistics. Int. J. Logist. Manag. **8**(1), 1–14 (1997)
7. Keely, L.C., Garcia-Dastugue, S.J., Lambert, D.M., Rogers, D.S.: The supply chain management processes. Int. J. Logist. Manag. **12**(2), 13–36 (2001)
8. Michael, K., McCathie, L.: The pros and cons of RFID in supply chain management. In: International Conference on Mobile Business (ICMB 2005), pp. 623–629. IEEE (2005)
9. Juels, A.: "Yoking-proofs" for RFID tags. In: Proceedings of the Second IEEE Annual Conference on Pervasive Computing and Communications Workshops, 2004, pp. 138–143. IEEE (2004)
10. Bolotnyy, L., Robins, G.: Generalized "yoking-proofs" for a group of RFID tags. In: 2006 Third Annual International Conference on Mobile, Ubiquitous Systems Networking & Services, pp. 1–4. IEEE (2006)
11. Ray, B.R., Abawajy, J., Chowdhury, M.: Scalable RFID security framework and protocol supporting Internet of Things. Comput. Netw. **67**, 89–103 (2014)
12. Song, B., Mitchell, C.J.: Scalable RFID security protocols supporting tag ownership transfer. Comput. Commun. **34**(4), 556–566 (2011)
13. Molnar, D., Wagner, D.: Privacy, security in library RFID: issues, practices, and architectures. In: Proceedings of the 11th ACM Conference on Computer and Communications Security, pp. 210–219. ACM (2004)
14. Fouladgar, S., Afifi, H.: Scalable privacy protecting scheme through distributed RFID tag identification. In: Proceedings of the Workshop on Applications of Private and Anonymous Communications, p. 3. ACM (2008)
15. Solanas, A., Domingo-Ferrer, J., Martínez-Ballesté, A., Daza, V.: A distributed architecture for scalable private RFID tag identification. Comput. Netw. **51**(9), 2268–2279 (2007)
16. Stickel, E.: A new method for exchanging secret keys. In: Third International Conference on Information Technology and Applications (ICITA 2005), vol. 2, pp. 426–430. IEEE (2005)

A Roadmap for Upgrading Unupgradable Legacy Processes in Inter-Organizational Middleware Systems

Radhouane B.N. Jrad[1(✉)], M. Daud Ahmed[2], and David Sundaram[3]

[1] OJI Fibre Solutions, Auckland, New Zealand
Rad.Jrad@OJIFS.co.nz
[2] Manukau Institute of Technology, Auckland, New Zealand
Daud.Ahmed@Manukau.ac.nz
[3] University of Auckland, Auckland, New Zealand
D.Sundaram@Auckland.ac.nz

Abstract. Complex changes in an Organization's Information Systems require roadmapping to ensure planning and execution meet the objectives. While traditional projects have plethora of methodologies to help achieving their goals, agile projects are harder to plan, particularly when dealing with Unupgradable Legacy Processes (ULP). A ULP is a process that is too old, complex, critical, and/or costly to be upgraded using standard methodologies and tools. One approach to address such difficulty is to separate project and technological roadmaps to separate focuses on organizational and technical aspects. In B2B context, and more precisely Inter-Organizational Information Systems (IOIS), the increasing need for integration has generated a new layer of middleware components referred to as Inter-Organizational Middleware Systems (IOMS). IOMS is a set of services, processes, procedures and methods that allow information to be shared between multiple partners of the same IOIS despite the heterogeneity of their systems. In spite of IOMS being relatively a new concept, it lacks full valuation and dare-we-say appreciation from stakeholders, which has ultimately culminated in them suffering the problem of ULPs. The purpose of this paper is to address the issue by proposing a set of roadmaps to upgrade ULPs in IOMS. First, the concept of roadmaps is investigated and a separation between Enterprise Project (EP) and Technological Project (TP) roadmaps is put forward. IOMS is then presented before a set of roadmaps is proposed to address its ULP issues. An implementation validating these roadmaps is then presented before merits and limitations of the proposed artifacts are discussed.

Keywords: Roadmap · Architecture · IOMS · IOIS · Middleware · Unupgradable · Legacy · Process · B2B · IADR · FUI · MAPIS · Action research · Design science

1 Introduction

A roadmap, in its broader sense, is a plan or a strategy aiming to achieve a specific predefined goal [1]. In organizational context, it can be regarded as a means for integrating technology and business towards developing technological strategies [2, 3]. As its name suggests, a roadmap is a visual or a descriptive construct (MAP) that illustrates

© Springer International Publishing AG 2016
R. Doss et al. (Eds.): FNSS 2016, CCIS 670, pp. 142–156, 2016.
DOI: 10.1007/978-3-319-48021-3_10

a methodological approach (ROAD) and specifies (i.e. guides) an explorer or a decision maker to conduct a journey or to devise a solution in a prescribed manner. While maps can offer alternative paths for the same journey, by nature they cannot be vague or inefficient. Roadmaps in Information Systems (IS) are often detailed and vary from being a simple representation of processes, e.g. cataloguing a software's release agenda, to a complex plan that aims to address a complex issue [4, 5]. They endeavor to answer the Why-What-How-When questions [6], link problems to solutions, and explain how elements of the system fit, interact and evolve together [7].

This paper attempts to address the lack of standards for upgrading legacy processes in Inter-Organizational Middleware Systems (IOMS) and proposes a set of IOMS-tailored roadmaps with different levels of abstraction. IOMS enables B2B process integration amongst partners of the same Inter-Organizational Information System (IOIS) [8]. Despite it being both a relatively-new and critical concept, IOMS is already suffering the prickly problem of legacy processes [9]. As a consequence, costs and efforts to maintain its processes' performance have become a known organizational problem. This research specifically endeavors to address the problem of Unupgradable Legacy Processes (ULP) in IOMS which are too old, complex, critical, and/or costly to be upgraded using standard methodologies and tools [8]. First, the concept of roadmaps is investigated, and the distinction between Enterprise Project and Technological Project roadmaps is detailed. IOMS is then presented with a focus on its ULPs and the difficulty to upgrade them. A set of roadmaps is then proposed to address the problem before an actual implementation validating them is presented. Finally, merits and limitations of the proposed artifacts are discussed.

2   Roadmaps

While roadmaps in IS serve to actually implement change, they are not loosely constructed. Instead, they are built above well-defined architectures and frameworks and supported by appropriate concepts and theories [7, 10]. However, even though they can be reused in the future [11], roadmaps' efficiency is often short-lived due to their heavy attachment to the current state of technology as well as existing business practices and conjuncture. They effectively deal with moving targets [12], and as such, they are more susceptible to obsoletion and loss of relevance than their underlying constructs like architectures. Figure 1 describes a general view of the relationship between building blocks in IS research and different investigations can select different elements from it. In particular, the justification of roadmaps for smaller changes can be challenged amid resource requirements. However, the intricacy and cross-disciplinary characters of emerging technologies, coupled with the uncertainties and risks they carry with them, bring significant complexity to IS. And it is such complexity that warrants the need of appropriate roadmaps [7, 11]. Indeed, they provide invaluable support to decision-making, performance improvement, and projects evaluation [11]. They also ensure inclusion of diverse perspectives as well as the acquirement of the all-important buy-in from various stakeholders [12]. In essence, a lack of proper project roadmap could potentially lead to project failure [11].

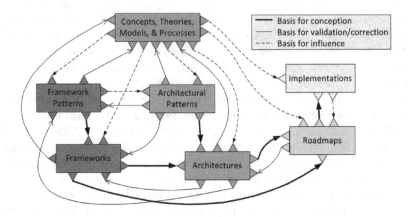

Fig. 1. Positioning Roadmaps in Research

There is a dual nature to roadmaps with them acting as forecast tool but also as action plans [3, 12]. Such duality is a strength by way of allowing the association of the Why-What-How-When inquiries, but also a challenge when dealing with unpredictable and disruptive changes [12]. A way to address these challenges is to separate roadmaps for projects execution from those for technology implementation. Such distinction allows constructing and assessing project steps without having to worry about challenges in technology deployment, and by the same token it allows planning for solutions implementation without focusing on business procedures and requirements for projects. The next 2 sections present the Enterprise and the Technological project roadmaps and detail their inner differences.

2.1   Enterprise Project (EP) Roadmaps

By definition, a project is an assembly of individuals collaborating towards a common goal [1]. Usually with an innovative spirit, projects are used to establish changes that otherwise cannot be promoted using standard processes [13]. They usually involve people from different disciplines and/or organizations [14], have a limited time frame, and their impact on the organization dictates how complex they can be [13]. Typically, there are two types of projects: Traditional and Agile. Traditional projects are well defined in scope, time, and responsibilities, and the end is fairly known before the start [15]. While these are useful in classical fields like manufacturing, they have well-documented limitations when they are applied in IS [16]. Indeed, IS projects tend to be complex with ill-specified goals, for which solutions are often not fully known at the outset of the project [17, 18]. Force-feeding inflexible PM methodologies into IS projects is in fact a step towards their failure [19, 20]. As a consequence, various Agile Project Management (APM) methodologies have been developed to address the issue [15–17, 19, 21].

While roadmapping for traditional projects is generally straightforward, trying to plan for an agile project is a serious challenge for the project itself [22], but also in reusing any existing roadmap. What was a good practice yesterday might be a legacy issue today. And the more detailed the roadmap the more time-specific it becomes. In a dynamic conjuncture, higher level roadmaps have better chances to be reusable. An even bigger challenge, however, is to plan for projects selection. Typically, an organization has a plan for multiple prospectus projects for one or many years and the task to order them in terms of criticality, importance, relevance, resources requirement, and need, is challenging. Because the selection of the wrong set of assignments locks valuable resources away on poor projects that would yield little results [23], the Enterprise Project Roadmaps need to also account for projects' selection and ordering.

2.2   Technological Project (TP) Roadmaps

Technological roadmapping is used to support organizational strategy and planning for technology and to help achieve the economic, social, and environmental goals associated with its use [4, 5]. It can be applied on products, services, and/or processes. As the complexity of technology increases, the cost of acquiring, managing, and updating it surges, and the need for effective processes to manage associated changes also increases [4]. Managing technology requires two important guarantees [5]: The constant and consistent linkage between technological resources and company objectives, and the existence of proper processes to manage technology.

While roadmaps for the EPs focus on dealing with the need to manage and implement technology, TP Roadmaps focus on the process of assessing what technology to select as well as where and how to implement it. The dynamic nature of IT poses a challenge of reusability for these roadmaps, and because of requirements for both field experience and roadmapping skills, the role of technological project planner can and should be regarded as an expert field of its own. Indeed, while the usability of previous roadmaps may be limited, the experience in developing them in a particular field increases with experience, but only for that specific field. It is therefore more beneficial from operations and cost perspectives to have technological roadmaps driven by internal resources. More specifically, every team in the organization's IS department should be the master of its roadmaps, albeit under the sponsorship of an experienced manager. Using external resources to develop technology roadmaps is less effective and costlier, particularly in the mid to long term. One obvious reason is that steps of problem and system discovery and assimilation would be repeated at the beginning of every project.

In their combined 30+ years dealing with IOIS and IOMS projects, one problem persistently noted by the authors is that both the EP and TP roadmaps can be mixed up or amalgamated. Often, most of what appears from the iceberg is technology itself, with business and organizational aspects of the project usually reduced in importance or assimilated as part of the technology upgrade. For instance, fixing a known network problem could erroneously be assumed to systematically improve business processes; getting vendor support for a project Go-Live could be falsely assumed as a support contract; even a simple database server version upgrade could wrongfully be thought of as an enhancement of business processes. As the next section will demonstrate, the argument to split EP and TP roadmaps takes even more meaning when the project involves middleware components in multi-organizational context.

3 Roadmaps for Upgrading ULPs in IOMS

3.1 Inter-organizational Middleware Systems (IOMS)

History suggests that since early human civilizations (Mesopotamia), doing business and making profits was always part of societal tissue. And the more advanced societies the more thriving the businesses and the more endorsed the concept of making profit [24]. While businesses were first focused on offering food, clothes, and shelter [24], a concept labeled nowadays as Business-to-Consumer (B2C), Business-to-Business transactions (B2B) started gaining interest when the principle of credits became an integral part of the business paradigm [24]. As means to amass wealth increased, businesses grew in sizes and ideas, but also in rivalry and complexity. Subsequently, they started to look at synergizing their efforts towards improving the management of costs [25] and competition [26]. With the arrival of IT, and more specifically Electronic Data Interchange standard (EDI), partnering organizations started integrating processes [27]. The change in shape of the B2B, however, did not stop there. Instead, it continued to morph reflecting the evolution of world's political and business configuration. Open online platforms have since become important drivers in the business world [28], relegating EDI to the rank of "Legacy" concepts. Undeniably, what started as mostly B2C [24] has now become an even share between B2B and B2C [29], and projections are for B2B volumes to continue increasing [28].

One of the key approaches to establishing successful B2B links is The Inter-Organizational Information System (IOIS). An IOIS is an automated IS shared and co-managed by participating organizations [30], each within their own business and technical contexts [20] (Fig. 2). IOIS are significantly complex [31, 32], and once integrated within an organization's systems, they are both immovably persistent and continuously evolving [33]. The concept of IOIS has existed for over 5 decades [16, 34] and has gathered interest from industries and researchers alike. However, while industries have endorsed it, it has failed to attract enough research [9, 34] despite it having links with virtually every major area of IS research [35]. In particular, handling IOIS diffusion over multiple organizations, locations, countries, legal systems, cultures, and time zones remains feebly investigated [16]. This is more so obvious when considering that despite the prosperity of Adaptive Business Networks (ABN) as a means to establish business partnerships [36], Enterprise Architectures have mostly regarded organizations as single entities [37]. Recently, however, a shift of interest in IOIS-related studies has been witnessed [38], and a prominent reason for it is that the concept has effectively moved from being a static system where partners, processes and roles are immobile, to being dynamic where alliances are continuously built and dismantled [34]. It is justifiably natural that researchers are more interested in studying dynamic phenomena than in re-studying static ones [36].

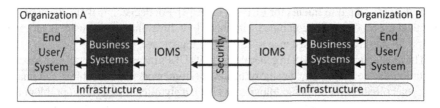

Fig. 2. Business process flow in inter-organizational information systems

The adaptive and dynamic aspects of IOIS are achieved through its flexibility to integrate and subtract partners from the business network quickly and affordably. The process is managed by the integrator component inside IOIS, referred to as the Inter-Organizational Middleware System (IOMS) [9]. IOMS is an inter-organizational collection of application and communication policies, procedures, methods, and services that allow partner's heterogeneous systems to communicate, exchange and validate business information [9] (Fig. 2). IOMS effectively enables the B2B process to cross organizational frontiers and resolves incompatibility issues between enterprise systems from different organizations utilizing different technologies [8].

Sitting in the middle, an IOMS is constantly pushed and pulled by external systems, including other IOMSs. As the complexity and agility of Enterprise Systems and IOISs increase, businesses' integration becomes more convoluted and complicated. With it, the use of IOMS can no longer be considered as optional. In effect, a new paradigm for business integration has already taken place creating the need for designing, architecting, governing, and managing IOMS on its specifities and merits, independently from other IOIS components and layers [36]. Notwithstanding such shift, reality remains that IOMS continues to be marginalized in enterprise strategies, architectures, and budgets, and only when business processes are affected that it gains attention. Subsequently, the relatively new concept has already witnessed the gargantuan problem of legacy processes [9]. As these legacy processes continue to fall behind technological advancement, they effectively become unupgradable [8], and common methodologies and tools are no longer effective in updating them [39]. In face of the problem, organizations resort to following the "least resistant path" by first trying to re-engineer the process, else trying to leave it unchanged, else to wrap it in more procedures. Factually, it is the process of wrapping that deepens the complexity of the black box the unupgradable processes have become [8]. Inevitably, businesses end up diverting extensive resources to keep these processes functional [9].

One of the reasons why IOMS is facing the ULP challenge is the fact it is hidden well inside the IOIS and presumed to be just another component of the system. This appears in the way research in general treats IOIS upgrades as one indivisible exercise, effectively excluding the possibility of addressing its components separately [8]. Furthermore, the lack of standards such as frameworks, architectures, and roadmaps specific to IOMS has been addressed with a relentless reliance on quick fixes, usually cheap but efficient workarounds. In the process, any existing system structure would be further damaged [40]. Moreover, while IOMS is part of IOIS, it is often considered as a sub-system of the Enterprise Resource Planning (ERP) system, and accordingly fits under the general requirements for ERP upgrades [8]. For decision makers, the complex technical aspect of IOMS, combined with its lack of direct impact on business turnover and profit, has created a resistance to address its legacy processes [9].

The management of IOMS in general, and of its ULPs in particular, requires a specific and tailored set of tools. The gavage of irrelevant frameworks, architectures, roadmaps, and project management methodologies has been a significant part of the problem and will only continue to exasperate it if not addressed. On the other hand, IOMS-specific concepts will not only address the problem of legacy processes, but also pave the road for more research to diversify and improve these concepts. In that context, the next three sub-sections will propose a set of roadmaps for managing the upgrade of ULPs in IOMS. The subsequent section will discuss their validation.

3.2   Meta Roadmap (MR) for Upgrading ULPs in IOMS

To address legacy processes in IOMS, organizations start with the use of tools and processes provided by the product vendor and the online community [8]. The availability of these solutions is tightly associated with the lifecycle of the IOMS product whereby every new version brings with it new requirements and guidelines including how to migrate from older versions. However, what would have been efficient in migrating legacy processes is less likely to work as the gap increases between the installed and the new versions. While it is often possible to migrate a current IOMS version to a newer platform, older versions may not be supported for such move, and a full replacement could instead be required. Though this problem can be identified in other IS fields, it is addressing it in the IOMS context that is lacking research. As a remedy, we propose a high level (Meta) roadmap to address ULPs in IOMS (Fig. 3). The Meta Roadmap (MR) has been constructed on top of the FUI framework [22, 40] and the MAPIS architecture [36], both of which are IOMS-specific.

The Meta Roadmap validates the argument that addressing ULPs requires both the EP and the TP roadmaps. On the other hand, it assumes that the baseline architecture and underlying framework used for the IOMS of the organization are IOMS-tolerant if not IOMS-specific. Furthermore, as IOMS holds business intelligence, business needs, impact and risks need to be accounted for when planning for IOMS changes, including for legacy process. Similarly to IOIS, IOMS is immovably persistent and continuously evolving; as such, changing IOMS products is neither an easy task nor a desirable decision. Owing to that, vendor requirements and guidelines have a significant influence on the organizational decision-making process associated with IOMS technology. In particular, the type and duration of licensing and support for a version of the IOMS solution can impact how the ULPs issue is approached. Indeed, longer lifecycles mean ULPs can be addressed in compact targeted tasks while shorter cycles mean they could be addressed as part of upgrade projects. Though it is factually challenging to unveil and understand the complexity of ULPs in the organization, fragmenting the Meta Roadmap into Enterprise Project and Technological Project roadmaps allows for addressing the matter in a structured and resilient way.

Fig. 3. Meta roadmap for upgrading unupgradable legacy processes in inter-organizational middleware systems

3.3 Enterprise Project Roadmap for Upgrading ULPs in IOMS

Figure 4 illustrates the proposed roadmap for Enterprise projects when addressing ULPs in IOMS. Addressing ULPS in reality an open-end journey, through which new legacy processes are involuntarily or carelessly engineered while others are being resolved. The continuation of the exercise depends on the organization's perception of the criticality and urgency associated with the ULP problem. Such judgments change over time and are not always in phase with reality. Only with more stakeholders' exposure to IOMS and its relevance to the business that support can be gained to manage and prevent legacy processes.

The upgrade of ULPs in IOMS relies on the use of Agile Project Management methodologies to account for unknowns encountered at the start and during the project [22] and to address multiple questions that may be unanswerable at the planning phase of the project. At the outset, the ULPs in IOMS need to first be identified and the depth and width of the ULP problem understood before undertaking any activities to address the situation. While in reality most ULPs are already known to the IOMS team, the formalization of the list sets the organization's appetite to (or not to) engage in addressing them. As per Fig. 4, the ULPs discovery would result in a set of Process lists (P-Lists) where they are ordered by priority based on their criticalness and (in)efficiency as well as the

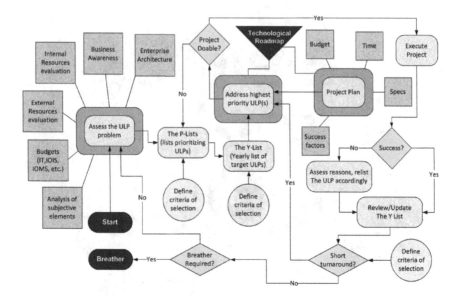

Fig. 4. Enterprise project roadmap for upgrading unupgradable legacy processes in inter-organizational middleware systems

associated impact/damage suffered by the business and the cost (effort and money) to keep them operational. The need for multiple P-Lists reflects the existence of different views and allows for different degrees of criticality. For instance, an organization might choose to create three lists of 'must-address-urgently', 'soon-to-become-a-problem' and 'nice-to-have'. At this stage, the complexity of the process redesign and reengineering is irrelevant. However, the organization should agree on a clear set of definitions to be used in ordering ULP processes.

Once established, the P-Lists are used to create the Yearly List (Y-List) which enumerates the processes selected for an upgrade in the next 12 months (or any equivalent period). At this point, the feasibility of migrating Y-List's ULPs is assessed. Once a candidate is selected, a project plan is prepared and the Technological Roadmap is drafted, all taking into account Change Management process and the business impact and risks associated with the change. If a candidate process is found to be tightly attached to one or more other ULPs, consideration needs to be given to migrating them at once. Mass-migration should however be discouraged unless unavoidable because of the costly and complex requirements it brings with it. That is in fact the very reason why the process became unupgradable in the first instance.

Once the project is completed, outcomes are assessed and the Y-List is updated. To allow for other activities to take precedence, the concept of a "Breather" is introduced in the roadmap. A Breather is a brief pause [1] that allows re-gathering focus and energy. Since this is not a project with a clearly marked end, it is important not to lock resources permanently or to put undue pressure on the organization. The Breather can be thought of as a timer before the project starts again, and its length is subjective. Mostly, it depends on the organization's perception of the ULP problem and the support from stakeholders to address it, both of which are inversely proportional to the gravity of the situation. Indeed, the more problematic the ULPs the stronger the desire to shorten the Breather, while the more manageable they are, the longer the Breather.

The proposed EP roadmap relies on sound P and Y Lists based on clear criteria definitions used in populating them. Such criteria are likely to change over longer periods of time particularly when the organization's structure changes. Events like acquisitions and divestitures have a significant impact on the structure of the business, its IS, the IOISs it subscribes to, and its IOMS.

3.4 Technological Project Roadmap for Upgrading ULPs in IOMS

While the EP Roadmap takes care of the approach to the ULPs from an organizational perspective, the Technological Project (TP) Roadmap looks at the actual upgrade process. It is technology-oriented and the more likely of the two roadmaps to suffer quick obsoletion. Usually, IS projects are approached in a linear fashion from the start to finish allowing them to encompass all required steps towards the project's targets. This is generally applied even to upgrading legacy processes. As such, the actual implementation of changes in IS projects is regarded as a set of tasks in the project plan that would be affected by the appropriate technical teams. However, addressing ULPs in IOMS is anything but linear and the Breather is a symptom of its nonlinearity. While standard IS projects usually have one project plan, different ULPs require different plans to address because each candidate has its own level of complexities requiring different resources and inputs from different teams. Furthermore, one single project plan cannot include information about all future iterations when they are still not investigated. The proposed TP Roadmap allows standardizing the approach while accounting for the cyclic need for project plans (Fig. 5).

As already stipulated, the TP roadmap is a component of the EP roadmap. It starts by assessing existing relationships between the candidate ULP and other business processes and potentially further ULPs. Efforts need to be put in documenting various processes and to engage various teams during the project, particularly when collecting technical input of how to reengineer the ULP. This in turn helps assessing project duration which would dictate the level of need for the involvement of other teams and third parties. Teams engagement should be promoted in a DevOps-like format where analysts, developers and operations personnel tightly interact and exchange knowledge [41]. The evaluation of internal support and resources is an important element of the roadmap because there is a need to guarantee their availability for the entire iteration of the ULP upgrade. Furthermore, the IOMS team being the driver for the project has a good understanding of the organization's structure and the strengths and weaknesses of other technical teams when it comes to supporting IOMS. Similarly, external support (vendor, consultants, etc.) also require evaluation and valuation.

For complex changes, a Proof-Of-Concept system (POC) might be required to improve understanding of required efforts and costs. A POC is a temporary system that is built specifically for the purpose of learning about a solution and its implementation processes in a lab-style environment [22]. All infrastructure requirements, including POC systems, need to be evaluated and their cost and deliverability

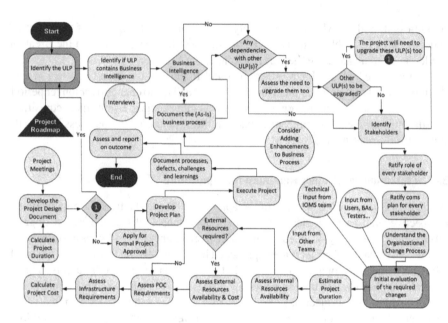

Fig. 5. Technological project roadmap for upgrading unupgradable legacy processes in inter-organizational middleware systems

determined. Once all details are gathered, the design, architecture, approval, execution, and reporting on the project can be performed as per standard processes.

3.5   Implementing the IOMS Roadmaps: A Case Study

The Meta, EP, and TP Roadmaps were implemented in a multinational organization with advanced e-commerce capabilities and belonging to multiple and disperse IOISs. The main project's aim was to upgrade the current IOMS to the latest version while accounting for legacy processes. ULPs had already been impacting on the business performance which served as the motivation to address them. The implementation, tests and validation processes followed the concepts and principles of Insider Action Design Research (IADR) which is suitable for investigations where applicability is as important as rigor and where feedback loops are part of the refinement process [9, 42]. The methodology allows the researcher to interact with the observed phenomenon while generating actual solutions as part of the research outcomes.

The previous IOMS upgrade was markedly expensive with ULPs being wrapped and exported as-is into the new version. These processes have subsequently remained unresolved and the organization was concerned that they would continue to drain significant resources. The idea to fully reengineer all ULPs at once was evaluated but discarded due to its high cost and risks. In reality, while IOMS upgrades should not cost excessively, ULPs drive prices and delays upwards requiring a lot of effort, time and even luck [8]. The organization nominated an employee from its IOMS team as the project driver and a squad was subsequently assembled from within the company.

First, the project team agreed on a framework and the baseline Architecture for the To-Be solution. This structural pre-project planning allowed the upgrade methodology to be consistent across IOMS processes and ensured that no new legacy processes were created as part of the project. It also allowed for the agreement to validate the roadmaps as a concept by including them in the project's Critical Success Factors.

The work started with the customization of the roadmaps. Eleven ULPs were identified and classified based on a combination of coefficients for business impact; system impact; risk of data loss; weakness of documentation; and complexity of reengineering. Each project team member used their judgment to attribute values for the coefficients after studying the ULPs. From the P-List only one candidate made it to the Y-List. The selected process was about using custom Java code used to connect to databases. The proposed solution was to upgrade processes away from this old method to instead use the DB Adapters offered by the vendor. The ULP was problematic because the Java service was written over the last 15 years and did not compile in the new version. Documentation was poor and some code was protected against access. Furthermore, this Java service was known to be sensitive to network glitches, forcing unplanned system restarts and undesirable outages. While keeping older version of Java and re-wrapping the service was an option to cheaply and quickly 'migrate' it, the organization was desperate to move away from such practice. To implement changes, the IOMS upgrade project represented a perfect opportunity since it already accounted for end-to-end testing. For prudence sake, however, it was decided not to upgrade more ULPs in the same project and instead concentrate on one critical ULP.

The Java DB connection process was analyzed and documented, including its inputs and outputs. A plan to reengineer it was drafted using feedback from IOMS developers as well as other teams including ERP, Database, Network, and System administrators. To refine the solution, the exercise took advantage of the POC system that was built for the upgrade project, and the duration, cost and effort estimates were produced. The implementation was completed in Development systems where unit testing was conducted, then in Test systems where integration and stress testing were completed, and then in Production systems. The fact that Test systems were true reflection of Production systems allowed testing to give confidence in the efficiency and stability of the solution before it going live. The upgrade of the ULP was successful, and while various stakeholders were being updated regularly, a final report was delivered with recommendation to move to the next ULP in 6 months after the upgrade. This Breather was introduced as a way to allow the new solution to be stable enough before starting modifying the system. Moreover, it ensured potential mix-ups between post-upgrade issues and ULP reengineering related problems are avoided.

4   Conclusion and Limitations

In this paper we proposed a set of roadmaps to address legacy processes in Inter-Organizational Middleware Systems that can no longer be addressed using standard methodologies and tools. These ULPs are haunting organizations because on one hand they do not fully understand and value the importance of IOMSs, and on the other hand they are forced to supply valuable resources in ensuring IOMS processes continue to perform. One of the reasons behind this dichotomy is that IOMS is relatively a new concept yet it lacks research and standardization. IOMS can no longer be addressed as part of the IOIS or the ERP systems, but instead it needs to be addressed as an independent concept with its specificities and requirements [36]. The development of the proposed roadmaps was designed to guide implementation of the FUI framework [22], the MAPIS architectural pattern [40] and the AIM Architecture [36], all of which have been constructed specifically for IOMS.

First, the Meta Roadmap was presented as a high-level view of how to approach upgrading the ULPs (Fig. 3). It highlights what needs to be addressed before being able to roadmap the actual changes. The Enterprise Project Roadmap was then proposed to look at the business aspect of the change (Fig. 4) by offering a methodology on what needs to be done to ensure the upgrade effectively resolves the organization's ULP issue while accounting for business rules and conjuncture. The Technological Project Roadmap was then proposed as a cascaded path to manage the actual upgrade of various ULP candidates (Fig. 5). It structures the implementation process and allows for a "Breather" to ensure organizational resources are not constantly consumed by the ULP issue. Finally, the set of artifacts was put to validation in the context of a large organization with an advanced IOMS trading in complex IOISs.

The next logical step would be to further explore the applicability of the roadmaps in different contexts such as in various legal frameworks or different industries. While they were constructed based on authors' prolonged experience with IOIS and IOMS projects, the artifacts in this paper have not been tested in the context of a medium or small sized organization. This limitation is driven by the assumption that the success in implementing the roadmaps in a complex scenario in an international organization with complex IOIS and IOMS processes would mean they are likely to succeed in simpler contexts. Not only this remains to be proven, but such limitation is in fact inherited from the underlying framework and architecture used in the project, and therefore needs focus if the roadmaps are to be generalized to medium and small size organizations. Further research could look at defining the application boundaries for the proposed roadmaps and potentially create sub-versions for different settings such as different industries, economical conjunctures, etc.

References

1. Oxford: Oxford Dictionaries (2014). http://www.oxforddictionaries.com
2. Groenveld, P.: Roadmapping integrates business and technology. Res. Technol. Manag. **50**(6), 49–58 (1997)
3. Probert, D., Radnor, M.: Technology roadmapping: frontier experiences from industry-academia consortia. Res. Technol. Manag. **46**(2), 26–30 (2003)
4. Radnor, D.R., Probert, M.: Technology roadmapping. Res. Manag. **47**(2), 24–37 (2004)
5. Phaal, R., Farrukh, C.J.P., Probert, D.R.: Technology roadmapping - a planning framework for evolution and revolution. Technol. Forecast. Soc. Chang. **71**(1–2), 5–26 (2004)
6. Albright, R.E.: A unifying architecture for roadmaps frames a value scorecard. In: IEMC 2003 Proceedings of the Managing Technologically Driven Organizations: The Human Side of Innovation and Change, pp. 383–386 (2003)

7. Albright, R.E.: Roadmapping convergence. In: Managing nano-bio-info-cogno innovations, vol. 2003, no. 12, pp. 23–31. Kluwer Academic Publishers, Dordrecht (2003)
8. Jrad, R.B.N., Ahmed, M.D., Sundaram, D.: Upgrading unupgradable middleware legacy processes: misconceptions, challenges, and a roadmap. In: Proceedings of the SEM 2013 1st International Workshop Software Evolution Modernization, vol. 1, no. 1, p. 8 (2013)
9. Jrad, R.B.N.: A roadmap for upgrading unupgradable legacy processes in inter-organizational middleware systems. In: 2014 IEEE Eighth International Conference Research Challenges Information Science, pp. 1–6, May 2014
10. Zachman, J.: A framework for information systems architecture. IBM Syst. J. **26**(3), 276–292 (1987)
11. Albright, R.E., Kappel, T.: How to use roadmapping for global platform products. PDMA Vis. **46**(2), 31–40 (2003)
12. Kappel, T.A.: Perspectives on roadmaps: how organizations talk about the future. J. Prod. Innov. Manag. **18**(1), 39–50 (2001)
13. Kuster, J., Huber, E., Lippmann, R., Schmid, A., Schneider, E., Witschi, U., Wüst, R.: Project Management Handbook. Springer, Heidelberg (2015)
14. Cleland, D.I., Ireland, L.R.: Project Management: Strategic Design and Implementation, 5th edn. McGraw-Hill Global Education Holding, New York (2007)
15. Kerzner, H., Belack, C.: Managing Complex Projects. Wiley, Hoboken (2010)
16. Jrad, R.B.N., Sundaram, D.: Inter-organizational information and middleware system projects: success, failure, complexity, and challenges. In: Americas Conference on Information Systems (AMCIS), 2015, p. 12 (2015)
17. Wysocki, R.K.: Effective Project Management: Traditional, Agile, Extreme, 7th edn. Wiley, Hoboken (2013)
18. Conforto, E.C., Salum, F., Amaral, D.C., da Silva, S.L., de Almeida, L.F.M.: Can agile project management be adopted by industries other than software development? Proj. Manag. J. **45**(3), 21–34 (2014)
19. Doherty, N.F., Ashurst, C., Peppard, J.: Factors affecting the successful realisation of benefits from systems development projects: findings from three case studies. J. Inf. Technol. **27**(1), 1–16 (2011)
20. Jrad, R.B.N., Sundaram, D.: Challenges of inter-organizational information and middleware system projects: agility, complexity, success, and failure. In: Proceedings of the Sixth International Conference Information, Intelligence Systems Application (IISA 2015), pp. 1–6 (2015)
21. Bosq, M.: The Inner Warrior: A Practical Guide to Fight Against Our Fears and to Conquer a Higher Level of Existence. iUniverse, Bloomington (2002)
22. Jrad, R.B.N., Sundaram, D.: Inter-organizational middleware systems: a framework for managing change. In: Proceedings of the Sixth International Conference Information, Intelligence Systems Applications (IISA) (2015)
23. Khalifa, Z., Burgan, M., Bregaj, T., Almallak, M.: Planning and Roadmapping Technological Innovations. Springer International Publishing, New York (2014)
24. Roberts, K.: The Origins of Business, Money, and Markets. Columbia University Press, New York (2011)
25. Nagy, A.: Difficulties in implementing the agile supply chain: lessons learned from interorganizational information systems adoption. In: Baskerville, R.L., Mathiassen, L., Pries-Heje, J., DeGross, J.I. (eds.) Business Agility and Information Technology Diffusion, pp. 157–171. IFIP International Federation for Information Processing, Atlanta (2005)
26. Chaffey, D.: Groupware, Workflow, and Intranets: Reengineering the Enterprise with Collaborative Software. Butterworth-Heinemann, Boston (1998)

27. Bussler, C.: B2B Integration: Concepts and Architecture. Springer, Heidelberg (2003)
28. The Global B2B E-commerce Market Will Reach 6.7 Trillion USD by 2020, Finds Frost & Sullivan (2015). http://ww2.frost.com/news/press-releases/global-b2b-e-commerce-market-will-reach-67-trillion-usd-2020-finds-frost-sullivan/. Accessed 29 Mar 2016
29. Lilien, G.L.: The B2B Knowledge Gap. Int. J. Res. Mark., no. May, February 2016
30. Hsu, P.F.: Integrating ERP and E-Business: resource complementarity in business value creation. Decis. Support Syst. **56**(1), 334–347 (2013)
31. Hekkala, R., Urquhart, C.: Everyday power struggles: living in an IOIS project. Eur. J. Inf. Syst. **22**(1), 76–94 (2012)
32. Eom, S.B.: An introduction to inter-organizational information systems with selected bibliography. In: Eom, S.B. (ed.) Inter-Organizational Information Systems in the Internet Age, pp. 1–30. IGI Global, Hershey (2005)
33. Reimers, K., Johnston, R.B., Klein, S.: An empirical evaluation of existing IS change theories for the case of IOIS evolution. Eur. J. Inf. Syst. **23**, 373–399 (2013)
34. Haki, M.K., Forte, M.W.: Inter-organizational information system architecture: a service-oriented approach. In: Camarinha-Matos, L.M., Boucher, X., Afsarmanesh, H. (eds.) Collaborative Networks for a Sustainable World. IFIP Advances in Information and Communication Technology, vol. 336, pp. 642–652. Springer, Heidelberg (2010)
35. Markus, M., Tanis, C.: The enterprise systems experience - from adoption to success. In: Zmud, R.W. (ed.) Framing the Domains of IT Management: Projecting the Future Through the Past, pp. 173–207. Pinnaflex Education Resources Inc., Cincinnati (2000)
36. Jrad, R.B.N., Sundaram, D.: Architecting adaptive inter-organizational middleware systems. In: Proceedings of the SAI Computing Conference 2016, p. 10 (2016)
37. Drews, P., Schirmer, I.: From enterprise architecture to business ecosystem architecture. In: 2014 IEEE 18th International Enterprise Distributed Object Computing Conference Workshops and Demonstrations, pp. 13–22 (2014)
38. Mueller, T., Schuldt, D., Sewald, B., Morisse, M., Petrikina, J.: Towards inter-organizational enterprise architecture management - applicability of TOGAF 9.1 for network organizations. In: American Conference on Information Systems 2013, At Chicago, USA, pp. 1–13 (2013)
39. Dumitraş, T., Narasimhan, P.: Why do upgrades fail and what can we do about it? Toward dependable, online upgrades in enterprise system. In: Bacon, J.M., Cooper, B.F. (eds.) Middleware 2009. LNCS, vol. 5896, pp. 349–372. Springer, Heidelberg (2009)
40. Jrad, R.B.N., Sundaram, D.: Architectural pattern for inter-organizational middleware systems. In: Proceedings of the International Conference on Nature of Computation and Communication (ICTCC) 2016, p. 6 (2016)
41. Smeds, J., Nybom, K., Porres, I.: DevOps: a definition and perceived adoption impediments. In: Lassenius, C., Dingsøyr, T., Paasivaara, M. (eds.) Agile Processes in Software Engineering and Extreme Programming. LNBIP, vol. 212, pp. 166–177. Springer, Heidelberg (2015)
42. Jrad, R.B.N., Ahmed, M.D., Sundaram, D.: Insider action design research: a multi-methodological information systems research approach. In: Research Challenges in Information Science (RCIS), pp. 28–30 (2014)

Sustainable, Holistic, Adaptable, Real-Time, and Precise (SHARP) Approach Towards Developing Health and Wellness Systems

Farhaan Mirza[1(✉)], Asfahaan Mirza[2], Claris Yee Seung Chung[2], and David Sundaram[2]

[1] Department of Information Technology and Software Engineering,
Auckland University of Technology, Auckland, New Zealand
farhaan.mirza@aut.ac.nz
[2] Department of Information Systems and Operations Management, University of Auckland,
Auckland, New Zealand
{a.mirza,claris.chung,d.sundaram}@auckland.ac.nz

Abstract. As populations age and chronic diseases become more prevalent, new strategies are required to help people live well. Traditional models of episodic health care will not be sufficient to meet changing health care needs and the reorientation of services towards maintaining function as opposed to treating illness. One strategy to meet these challenges is an increased focus on self-care via use of broader social networks and seamless integration of applications with lifestyle activities, particularly for people with chronic diseases including diabetes, cardiovascular disease, and respiratory conditions. There has also been a rapid increase in a range of technologies for connecting different components of the health system and delivering services through smartphones and connected devices. Our proposal is to pursue systems development in healthcare in a way that considers a range of aspects known as SHARP: Sustainable, Holistic, Adaptive, Real-time and Precise. This approach will provide solutions that will be useful and effective for managing the long-term well-being of individuals.

Keywords: Sustainable health systems · Precision health · Disease management · Adaptive health systems · Self-managed healthcare applications

1 Introduction

As populations age and chronic diseases become more prevalent, new strategies are required to help people live well. Traditional models of episodic health care will not be sufficient to meet changing health care needs and the reorientation of services towards maintaining function as opposed to treating illness.

One strategy to meet these challenges is an increased focus on self-care and the use of broader social networks and integration of technologies and seamless applications to support health and wellness activities of individuals, particularly for people with diabetes, cardiovascular disease (ischemic heart disease and heart failure), and respiratory conditions including asthma and Chronic Obstructive Pulmonary Disease (COPD). In New Zealand, the Health Research Council (HRC) has issued a 2016 health and

© Springer International Publishing AG 2016
R. Doss et al. (Eds.): FNSS 2016, CCIS 670, pp. 157–171, 2016.
DOI: 10.1007/978-3-319-48021-3_11

wellbeing research investment signal with a specific focus on keeping people healthy and independent throughout life [1]. In the last 18 months, the Ministry of Health in New Zealand has heavily promoted the use of *patient portals* to improve patient engagement with their own care, and has just funded a training program to help primary care support self-management, primarily focused on diabetes care [2]. Another strategy is the refocusing of health systems, particularly the UK NHS, on the experience of the *health consumer*, measuring patient experience both as a mechanism to tune systems to better meet need, and also as a primary outcome [3]. Against the background of these philosophical shifts, there has been a rapid increase in a range of technologies for connecting different components of the health system and delivering services through smartphones and connected devices.

Our proposal is to encourage *systems development in healthcare* in a way that considers a range of aspects we named SHARP: Sustainable, Holistic, Adaptive, Real-time, and Precise. This approach will provide solutions that will be useful and effective for managing long-term wellbeing of individuals.

1.1 Motivation and Gap

The motivation for this work stems from the following challenges in healthcare around the world.

- A*geing populations*, which consume a greater amount of healthcare services and have limited tax-raising ability.
- *Longer life expectancy* of individuals [4], and this involves *long-term disease management*, which changes the paradigm of treatment from episodic to ongoing.
- *Disadvantaged communities.* For instance, in New Zealand the type-two diabetes prevalence is seven percent, with higher rates among Maori (9.8 %), and Pacific peoples (15.4 %) [5].
- *Changing lifestyle patterns* of single parents, smaller families, and unhealthy lifestyle factors particularly smoking, physical inactivity, poor diet quality, and stagnant or worsening rates of obesity [6].
- *Raised patients' expectations* of healthcare to be delivered in a way that leverages digital devices and web technologies. For example, the authors in [7] state that shared medical records including Internet-accessible records are almost universally endorsed across a broad range of ethnic and socioeconomic groups.

The gap occurs because current health programs, initiatives, systems, and applications focus on specific outcomes of medical treatment, awareness, and communication between patient and health services. The treatment from a clinician-patient interaction perspective is often episodic, meaning the treatment plan is adjusted every time the patient is seen by a medical professional. After each episode, patients find it challenging to adhere to recommended self-management programs, particularly the adoption of different lifestyle patterns. During the time the patient lives independently (which is desired) away from health services, the mismanagement of diet, exercise, and medication adherence contributes towards unfavorable health outcomes.

To develop systems and processes that can sustain long-term success and appreciate the holistic nature of healthcare delivery, we propose development of solutions that are standardized and integrated across a full range of stakeholders. The following section presents the SHARP approach to build and evaluate successful eHealth projects.

2 SHARP Approach

The key research question is how do we care for the wellbeing of our ageing, yet-to-age, and individuals with chronic illnesses in a holistic and sustainable manner? This study aims to provoke research and develop projects that implement and evaluate a SHARP approach for management of chronic diseases to support and enhance the well-being of the population.

The SHARP approach is both a concept and a process (Fig. 1). We believe that having a sustainable vision for society and the individuals and communities within it is at the heart of the approach as well as its starting point. To carry out this vision, we need to have a holistic perspective that spans the multiple dimensions of the individual. Furthermore, we need to build within the people, processes, and systems the ability to be adaptive and agile, and to learn, and respond. This can only be accomplished if we have access to real-time information — information that is available to different stakeholders at the right time, right place, and appropriately personalized so they can make quick and effective decisions based on scientific evidence derived from trending historical data. The afore-mentioned real-time data will be captured via devices and sensors and rigorous analysis. It is this type of personalized, preventative, predictive, patient-centered, and precise information that will result in solutions to support health goals of individuals, communities, and society. The following sections articulate the SHARP concepts and their impact on the development of eHealth systems.

Fig. 1. SHARP approach and process

2.1 Motivation and Gap

Due to an ageing population, there is growing interest in long-term sustainable health management and associated systems. This means the system should be sustainable and support patients' sustainability. The focus of individual and family sustainability is on economic, environmental, social, health, and cultural aspects.

Balancing the various dimensions of life (Fig. 2) is a challenging task — especially for ageing and chronic condition patients. Unlike acute diseases, the duration of chronic conditions is often life-long, and patients often face life uncertainties with a great amount of anxiety. Their main concerns are the health issues; however, lost productivity due to their illness and the costs of healthcare easily hurt the financial status of patients and their families and make them more vulnerable to impoverishment [8]. Furthermore, this stress can impact their social and family relationships, which can make their health condition worse.

Fig. 2. Sustainability dimensions

The threat of patients' sustainability causes a huge *cost of illness* (COI) burden to our society. For example, the US economy spends $1 trillion annually for the most common chronic illness healthcare [9]. Also, US employers bear a cost of $1,685 per employee annually because of health-related absenteeism [10]. Fortunately, much of these costs are avoidable by promoting and supporting sustainable life patterns for chronic illness patients, and by preventing those who are healthy from falling ill to diseases that could be avoided. Therefore, it is important to provide a health system that can understand an individual's life and support them in a sustainable manner in the different facets of their life: *health, economic, environmental, social*, and *cultural*.

SHARP takes sustainability as one of the main concepts and provides insightful information to segments such as ageing, yet-to-age, individuals with chronic illnesses, disadvantaged communities, and patients who have raised expectations of healthcare. As SHARP understands dimensions of individual sustainability (Fig. 3), it will consider

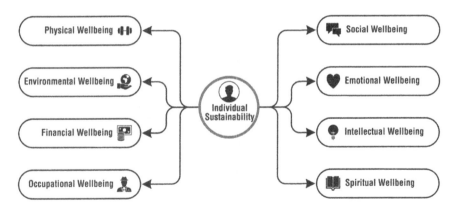

Fig. 3. Factors of individual sustainability and wellbeing

various risk factors related to our life and provide sustainable health advice by connecting individuals, families, and various health providers.

2.2   Holistic

Our life has various dimensions that are tightly intertwined [11]. Understanding their relationship is important to the pursuit of a sustainable life. Dunn [12] asserted the definition of health should not be limited to an absence of illness, but rather looked at as the *integrated being* of an individual that includes body, mind, and spirit.

As such, a health care system should be able to offer an integrated and holistic care approach. *Integrated care* means health systems not only provide clinical treatments, but also take care of the full range of patients' needs from the physiological, spiritual, emotional, and neurological dimensions (Fig. 4). One of the health promotion campaigns in the US adopted a holistic approach to life in their treatment of mentally ill patients; their campaign believes this is the only way to deliver a high-level of wellbeing to patients [13].

As patients are already experiencing some physical limitations, the holistic approach to life can reduce anxiety arising from life's uncertainties.

An effective management of health conditions should extend beyond medical treatment. Therefore, SHARP caters to life dimensions that are likely to impact on one's health condition, and vice versa. It will educate patients and their families on relationships and dynamics of life dimensions by picturing patients' lives as scenarios. These scenarios embrace the concept of the interrelationship of life's dimensions in creating individual sustainability, and thereby help patients learn how one small action can transform the sustainability of their life.

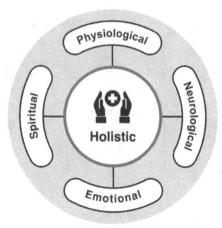

Fig. 4. Factors of holism in SHARP

2.3 Adaptive

Adapt or die is a common phrase one often hears; however, the *way* one adapts is not always clear. When one looks at literature, there are many synonyms as well as phrases that help us to understand what adaptation is, and what is required to adapt — such as learning, being agile, versatile, alive, resilient, and independent (Fig. 5). While these characteristics of adaptation are vital, the process of adaptation is also important.

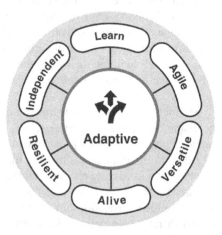

Fig. 5. Factors of adaptive in SHARP

Many processes such as sense-interpret-respond, sense-interpret-decide-act, observe-orient-decide-act, define-measure-analyze-improve-control, and strategize-design-implement-control, have been suggested in literature. At the heart of all of these is the requirement to sense-interpret-decide-act. The SHARP approach and process embodies, as a whole, an adaptive philosophy that helps the individual, the systems, and the support community around the individual to be constantly learning, alive, and resilient through agility, versatility, and independence.

2.4 Real-Time

A key aspect involving critical or important events is timing. For example, in the airline industry, *timing* is clearly defined: time to check in, time to board, and time to take off. However, the notion of real-time health care services is lacking. The model of healthcare services is designed to be an *episodic intervention* between the individual and health care sector. The individual is expected to visit the doctor when required; i.e., general checkup, vaccination, when sick, and regularly managing long-term disease. The individual is also expected to seek help when it is required; i.e., calling the ambulance, visiting the emergency department, or visiting a specialist.

The SHARP approach involves the use of real-time health services, which are a means of delivering a health intervention at the right time. The SHARP framework considers data latency, analysis latency, and decision latency (Fig. 6). Knowing these parameters and establishing the connectivity amongst stakeholders will enable delivery of the appropriate real-time services. For example, if a system knows the *data* (prescription information of a patient), it will be able to analyze *when* the prescribed *activity* is expected, and as a result, a *decision* might be to deliver a notification using suitable Ubiquitous Information System (UIS) media (see Sect. 2.5).

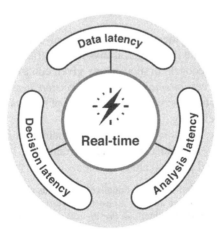

Fig. 6. Factors of real-time in SHARP

UIS technology is useful in delivering real-time contextual health alerts or feedback; many studies often rely on mobile technology or UIS to deliver real-time context-driven health services [14–16]. UIS could also help operationalize decision support objectives for complex disease management such as with diabetes. The study [17] shows a way to manage diabetes that uses an Internet of Things (IoT) approach.

Incorporation of the real-time part of the SHARP framework leads to achieving a proactive non-episodic model of health interventions. The real-time health alerts and feedback could assist in reminding patients of medication and health activities.

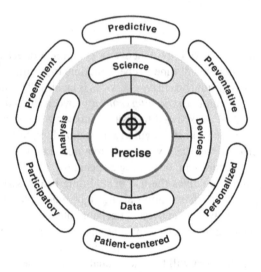

Fig. 7. Factors of precise in SHARP

2.5   Precise

We define *precision health* as the ability to deliver real-time feedback and support for health services tailored to the individual; it involves determining the most appropriate intervention, reminder, feedback, or alert in any given context. There is a variation of terminology between *precision medicine* and *precision health: precision medicine* focuses on clinical contributions; whereas, the notion of *precision health* is broader and focuses on individual wellbeing.

New global initiatives such as [18, 19] are advancing individualized patient care, and extend into university research centers and programs [20, 21], as well as calling for funding proposals [22] and political announcements:

> "I'm launching a new Precision Medicine Initiative to bring us closer to curing diseases like cancer and diabetes — and to give all of us access to the personalized information we need to keep ourselves and our families healthier" — *President Barack Obama [23].*

The notion of precision medicine or health can be further explored from *four* key aspects of the SHARP perspective (see Fig. 7). Firstly, it involves *science,* which is related to the medicine or clinical approaches. This is where studies involve clinical contributions; for example, in [24] the authors apply precision health care in a therapeutic approach to prevention in children's mental health, [25], look at advancing precision medicine for large-scale cancer genomics data, and [26] explain how precision medicine is useful for complex diseases including cancer, genetic disorders, and analysis of biomedical information including molecular, genomic, cellular, clinical, behavioral, physiological, and environmental parameters.

Secondly, it involves *devices* that use a range of UIS devices and services including mobile devices, mobile applications, web applications, tablet applications, Short Message Service (SMS), Email, mobile notifications, chat and instant messaging, smart watches and health bands, intelligent clothing accessories, IoT sensors and micro controllers, web-based widgets and modules, dashboards, kiosks, and screens. This wide range of channels enables easy delivery of the context-driven, precision health outcomes.

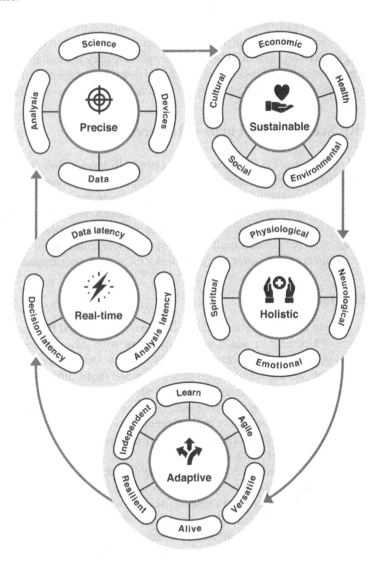

Fig. 8. The SHARP framework

Thirdly, it involves *data,* which is the ability to have *access* to the wide range of sources of data. Each individual's health data can be categorized into bio-molecular data, environmental data, and device data [27]. The bio-molecular data comprises data from exposome, epigenome, microbiome, metabolome, proteome, transcriptome, and genome. The environmental data includes medical, healthcare, biometric information, and scientific records. The device gathers data from wearables, mobile phones, IoT, cognitive analytical platforms like IBM Watson, and other cloud-computing health repositories.

Fourthly, precision health involves *analysis* which reveals insights, patterns, and relationships across data in an effort to deliver the right treatment to the right patient at the right time through early diagnosis and individually tailored treatments.

The above factors can be helpful in achieving *Precision Health* goals that (a) attempt to predict and proactively prevent ill health (b) are participatory and collaborative, and (c) provide personalized patient-centered health care that is pre-eminent [20] and inter-weaves all the ideas presented in Sect. 2 results in the SHARP Framework (Fig. 8).

3   SHARP Scenarios for Further Research

An estimated one in seven US adults have at least two of five major chronic conditions: cardiovascular disease, cancer, chronic obstructive pulmonary disease, diabetes, and arthritis [28]. Based on the literature in Sect. 2, the article presented how SHARP framework's characteristics can inform, design, and evaluate a comprehensive solution. Healthcare delivery on an ongoing basis based upon the SHARP approach can have a positive impact on an individual's health. A few specific scenarios that are being investigated by the authors of this article using the SHARP approach are presented below.

3.1   Medical Adherence

A shocking statistic published by the Center for Disease Control and Prevention in the US reports that 75 percent of adults are non-adherent with their medications [29]; furthermore, 31 percent of the patients do not adhere to new medications. On the other hand, 89 percent of patients acknowledge their prescribed medication and its medical adherence was necessary for maintaining health [30]. The research [31] reported the estimated annual costs resulting from poor medical adherence is $100 billion in the US and €25 billion in the European Union. Non-adherence can also lead to severe outcomes including deaths; it is estimated that non-adherence to antihypertensive treatment causes 89,000 premature deaths in the US annually [32].

At present, there are many methods for assessing patients' medical adherence [33], identified as: individual self-report of medication taking, real-time electronic monitoring, pharmacy refills, prescription claims databases, and pill counts. These are the most commonly used measures of adherence in research. However, there is no global or national standard for measuring medical adherence [34].

Based on the dreadful statistics above, and the development and adoption of UIS, we are presented with a timely opportunity to contribute solutions. Using the SHARP approach, the opportunity would first be to remind patients by delivering precise medical adherence interventions, and second, to obtain feedback on whether the medication taken was effective (a step beyond the precise medical adherence reminders). This feedback can be received via a feedback request, and delivered via the most suitable UIS medium. Finally, we can allow the ability to respond to or escalate negative feedback should the medication cause unfavorable results including allergic reactions. The response can be one that is clinical decision supported and delivered in real-time using bots, or involve a clinical intervention depending on the configured scenarios.

3.2 Diabetes Therapy for Ageing Individuals in Ambient Assisted Living (AAL) Environments

The prevalence of diabetes and impaired glucose tolerance is found to be similar for men and women, but increased with age [35]. Diabetes therapy management in Ambient Assisted Living (AAL) environments such as old people's and diabetes patient's homes, is a difficult task because it involves several factors that affect a patient's blood sugar levels [17]. Factors such as illness, treatments, physical and psychological stress, physical activity, drugs, intravenous fluids, and change in the meal plan can cause unpredictable and potentially dangerous fluctuations in blood sugar levels. However, diabetes therapy and treatments provided by physicians to the patients *do not* consider these fluctuations in real-time; this results in dosage inaccuracies in medication and may culminate in hyperglycemia and hypoglycemia episodes [17].

Using a SHARP approach, the development of an artificial intelligence (AI)-based clinical decision support mechanism can assist diabetic individuals with effective decision support. Furthermore, it can recommend precise insulin infusion calculations via an analytics-based intelligent decision support system. This could potentially assist patients with diabetes to regulate the amount of insulin introduced in accordance with their lifestyle and individual inputs. Consequently, it can lead to adjustments in software predictions of insulin requirements, and thus deliver personalized medicine to manage insulin levels for people who are unable to self-manage their disease in AAL.

3.3 Seamless Monitoring in Ambient Assisted Living (AAL)

The role of UIS in AAL environments presents us with an opportunity to achieve a seamless integration of connected sensors and cloud-based data collection possibilities. Using IoT in the personal home environment or care facilities can broaden the sources of *data* that can inform precision health scenarios. Automatic semantic summarization of human activity and detection of unusual inactivity are useful goals for a vision system operating in a supportive home environment [36]. For example, using a weight-sensor configuration embedded in a bed can be helpful in monitoring data and detecting early indications of physical or mental health changes. Other examples include tracking blood pressure, movement activity, or fall detection. AAL highlights technologies that adapt to the user rather than requiring the user to adapt to the technology [37].

Low-cost *appcessories* (including mobile application and sensors) and cloud-based services can be developed to support patients who want to remain in an independent living environment. It also avoids higher medical costs if conditions are detected at an early stage.

We believe the seamless monitoring approach will be successful as it improves the patient's experience of living independently, and at the same time includes relatives and caregivers delivering a low cost option of a *real-time care solution*.

3.4   Ageing Individuals

Healthy life expectancy increases across countries at a faster rate than total life expectancy, which suggests reductions in mortality are accompanied by reductions in disability [38]. This further suggests that a focus on the reduction of disability due to illness will become a key success factor in aged care and introduces new challenges for ageing populations.

The *new old* have a desire to live independently as long as possible. Since the traditional model of living closely together as a family is vanishing, and assisted care options like rest homes, hospitalization, institutionalization, and retirement villages are expensive options, the ageing individuals can continue to live in AAL with the assistance of *smart* devices. The authors [39] state the concept of the *smart home* is a promising and cost-effective way to improve home care for the elderly and disabled in a non-obtrusive way, and allows greater independence, maintenance of good health, and prevention of social isolation.

Generally, this population represents the future users of the *smart home* approach, provided the devices are effective and easy to use. Ageing individuals are often challenged with long-term conditions, so self-management becomes vital. [40].

The inclusion of multiple care groups; i.e., family, primary care, and secondary hospital, and including the contexts of work, home, and social, will allow development of successful solutions.

3.5   Digital Natives

The authors in [41] present a research commentary on digital natives, claiming their rise, along with the growth of UIS, potentially represents a fundamental shift in the paradigm for information systems research. To manage individuals with fewer health issues who are well-acquainted with UIS is an attractive opportunity to introduce preventative medicine health solutions. The author in [42] presents ways that UIS can expand the range and flexibility of intervention and teaching options available in preventive medicine and healthcare.

UIS can enable the delivery of automated, self-instructional, health behavior change programs through the Internet. Digital natives, having no adoption barriers, can be the biggest benefactors of such solutions. The opportunities for a SHARP-oriented approach with digital natives can involve studies on automated data collection, convenience healthcare, open communication platforms comprising social networks and video, and creation of *information environments* via digital platforms. The focus must extend beyond home and social to also include workplace wellness [43]. Overall, there is some evidence from physical activity self-help interventions that suggests enhancing the interactivity of intervention delivery increases the sustainability of intervention effects [42].

4 Conclusion

This review argues that traditional models of episodic health care will not be sufficient to meet changing health care needs. Maintaining an individual's health and wellbeing does not solely rely on advancement of clinical treatments and interventions. While these are important, there is now broad consensus that health differences between groups and within groups are not driven by clinical care, but by the social-structural factors that shape our lives [44].

This review presents an approach to pursue systems development in healthcare in a way that considers a range of aspects known as SHARP: Sustainable, Holistic, Adaptive, Real-time and Precise. This approach provides solutions that will be useful and effective for managing the long-term well-being of individuals. Furthermore, potential healthcare *scenarios* and *how* they can apply SHARP are presented. These scenarios include SHARP solutions for medical adherence, diabetes therapy, seamless monitoring, ageing individuals, and digital natives.

References

1. Health Research Council of New Zealand: HRC Investment Signals 2016 - General Guidelines. HRC, Auckland (2016)
2. Ministry of Health: Sharing Health Information: Toward Better, Safer Care. Ministry of Health, Wellington (2013)
3. Bate, P., Robert, G.: Bringing User Experience to Healthcare Improvement: The Concepts, Methods and Practices of Experience-Based Design. Radcliffe Publishing, London (2007)
4. Oeppen, J., Vaupel, J.W.: Broken limits to life expectancy. Science **296**, 1029–1031 (2002)
5. Krebs, J.D., Parry-Strong, A., Gamble, E., McBain, L., Bingham, L.J., Dutton, E.S., Tapu-Ta'Ala, S., Howells, J., Metekingi, H., Smith, R.B.W., Coppell, K.J.: A structured, group-based diabetes self-management education (DSME) programme for people, families and whanau with type 2 diabetes (T2DM) in New Zealand: an observational study. Prim. Care Diabetes. **7**, 151–158 (2013)
6. Ma, J., Rosas, L.G., Lv, N.: Precision lifestyle medicine: a new frontier in the science of behavior change and population health. Am. J. Prev. Med. **50**, 395 (2016)
7. Ross, S.E., Todd, J., Moore, L.A., Beaty, B.L., Wittevrongel, L., Lin, C.-T.: Expectations of patients and physicians regarding patient-accessible medical records. J. Med. Internet Res. **7**, e13 (2005)
8. Kankeu, H., Saksena, P., Xu, K., Evans, D.B.: The financial burden from non-communicable diseases in low- and middle-income countries: a literature review. Health Res. Policy Syst. **11**, 31 (2013)
9. Devol, R., Bedroussian, A.: an unhealthy america: the economic burden of chronic disease. Stroke (2007)
10. Stewart, W.F., Ricci, J.A., Chee, E., Morganstein, D.: Lost productive work time costs from health conditions in the United States: results from the American Productivity Audit. J. Occup. Environ. Med. **45**, 1234–1246 (2003)

11. Pukeliene, V., Starkauskiene, V.: Quality of life: factors determining its measurement complexity. Eng. Econ. **22**, 147–156 (2011)
12. Dunn, H.L.: What high-level wellness means. Can. J. Public Health **50**, 447–457 (1959)
13. Swarbrick, M.A.: Integrated care: wellness-oriented peer approaches: a key ingredient for integrated care. Psychiatr. Serv. **64**, 723–726 (2013)
14. Morris, S., Paradiso, J.: Shoe-integrated sensor system for wireless gait analysis and real-time feedback. In: Engineering in Medicine and Biology, 2002. 24th Annual Coference and the Annual Fall Meeting of the Biomedical Engineering Society EMBS/BMES Conference, 2002. Proceedings of the Second Joint (2002)
15. Wing, C., Yang, H.: FitYou: integrating health profiles to real-time contextual suggestion. In: Proceedings of the 37th International ACM SIGIR Conference on Research and Development in Information Retrieval - SIGIR 2014, pp. 1263–1264 (2014)
16. Boyes, A., Newell, S., Girgis, A., McElduff, P.: Does routine assessment and real-time feedback improve cancer patients' psychosocial well-being? Eur. J. Cancer Care (Engl).**15**, (2006)
17. Jara, A.J., Zamora, M.A., Skarmeta, A.F.G.: An internet of things–based personal device for diabetes therapy management in ambient assisted living (AAL). Pers. Ubiquitous Comput. **15**, 431–440 (2011)
18. Dankwa-Mullan, I., Bull, J., Sy, F.: Precision medicine and health disparities: advancing the science of individualizing patient care. Am. J. Public Health **105**, S368–S368 (2015)
19. Comings & Goings, JIM 63–8. J. Investig. Med. **63**, 893–897 (2015)
20. Precision Health. http://med.stanford.edu/precisionhealth.html
21. Precision Medicine: delivering the right treatment to the right patient at the right time through early diagnosis and individually tailored treatments. http://precisionhealth.uahs.arizona.edu
22. Empowering data-driven health. http://www.precisiondrivenhealth.com
23. Fox, J.L.: Obama catapults patient-empowered Precision Medicine. Nat. Biotechnol. **33**, 325 (2015)
24. August, G., Cicchetti, D., Gewirtz, A.: Moving toward precision healthcare in children's mental health: new perspectives, methodologies, and technologies in therapeutics and prevention. Dev. Psychopathol. **28**, 889 (2016)
25. Cully, M.: Anticancer drugs: advancing precision medicine in silico. Nat. Rev. Drug Discov. **14**, 311 (2015)
26. Collins, F.S., Varmus, H.: A new initiative on precision medicine (2015). http://dx.doi.org.ezproxy.aut.ac.nz/10.1056/NEJMp1500523
27. Harrer, S.: Measuring life: sensors and analytics for precision medicine, 1 June 2015
28. Ford, E.S., Croft, J.B., Posner, S.F., Goodman, R.A., Giles, W.H.: Co-occurrence of leading lifestyle-related chronic conditions among adults in the United States, 2002-2009. Prev. Chronic Dis. **10**, 120316 (2013)
29. Cohen, R.A., Villarroel, M.A.: Strategies used by adults to reduce their prescription drug costs: United States, 2013. NCHS Data Brief. vol. 184, pp. 1–8 (2015)
30. Horne, R., Weinman, J.: Patients' beliefs about prescribed medicines and their role in adherence to treatment in chronic physical illness. J. Psychosom. Res. **47**, 555–567 (1999)
31. Conn, V.S., Ruppar, T.M., Enriquez, M., Cooper, P.: Medication adherence interventions that target subjects with adherence problems: systematic review and meta-analysis. Res. Soc. Adm. Pharm. **12**, 218–246 (2016)
32. Beni, J.B.: Technology and the healthcare system: implications for patient adherence. Int. J. Electron. Healthc. **6**, 117 (2011)
33. Atreja, A., Bellam, N., Levy, S.R.: Strategies to enhance patient adherence: making it simple. MedGenMed **7**, 4 (2005)

34. Lehmann, A., Aslani, P., Ahmed, R., Celio, J., Gauchet, A., Bedouch, P., Bugnon, O., Allenet, B., Schneider, M.P.: Assessing medication adherence: options to consider. Int. J. Clin. Pharm. **36**, 55–69 (2014)

35. Scragg, R., Baker, J., Metcalf, P., Dryson, E.: Prevalence of diabetes mellitus and impaired glucose tolerance in a New Zealand multiracial workforce. N. Z. Med. J. **104**, 395–397 (1991)

36. Nait-Charif, H., McKenna, S.J.: Activity summarisation and fall detection in a supportive home environment. In: Proceedings of the 17th International Conference on Pattern Recognition, 2004. ICPR 2004, vol. 4, pp. 323–326. IEEE (2004)

37. Garcia, N.M., Rodrigues, J.J.P.C.: Ambient Assisted Living. CRC Press, Boca Raton (2015)

38. Mathers, C.D., Sadana, R., Salomon, J.A., Murray, C.J., Lopez, A.D.: Healthy life expectancy in 191 countries, 1999. Lancet **357**, 1685–1691 (2001)

39. Chan, M., Campo, E., Estève, D., Fourniols, J.-Y.: Smart homes — current features and future perspectives. Maturitas **64**, 90–97 (2009)

40. Tricco, A.C., Ivers, N.M., Grimshaw, J.M., Moher, D., Turner, L., Galipeau, J., Halperin, I., Vachon, B., Ramsay, T., Manns, B., Tonelli, M., Shojania, K.: Effectiveness of quality improvement strategies on the management of diabetes: a systematic review and meta-analysis. Lancet **379**, 2252–2261 (2012)

41. Vodanovich, S., Sundaram, D., Myers, M.: Research commentary —digital natives and ubiquitous information systems. Inf. Syst. Res. **21**, 711–723 (2010)

42. Fotheringham, M.J., Owies, D., Leslie, E., Owen, N.: Interactive health communication in preventive medicine: Internet-based strategies in teaching and research. Am. J. Prev. Med. **19**, 113–120 (2000)

43. Loeppke, R., Edington, D., Bender, J., Reynolds, A.: The association of technology in a workplace wellness program with health risk factor reduction. J. Occup. Environ. Med. **55**, 259–264 (2013)

44. Bayer, R., Galea, S.: Public health in the precision-medicine era. N. Engl. J. Med. **373**, 499–501 (2015)

Connected Bicycles

Otto B. Piramuthu[✉]

Buchholz High School, Gainesville, FL 32606, USA
obpira@gmail.com

Abstract. As IoT (Internet of Things) applications pervade every facet of our lives, it becomes necessary to take stock of the possibilities that include what has already been achieved and what could readily be achieved. We consider a specific facet of IoT applications as they relate to bicycles, specifically the use of IoT in connected bicycles. We discuss current IoT applications in connected bicycles as well as associated dimensions on connected and quantified self. While the concept of quantified self existed without any influence from IoT, the widespread acceptance of IoT and associated convenience have certainly spurred the emergence of IoT-enabled devices that facilitate ease of quantified self data collection. We also identify possible extensions to what already exists in connected bicycles from an IoT-based perspective.

Keywords: Connected bicycle · IoT · Quantified self

1 Introduction

The increase in awareness of the deleterious environmental effects of automobiles and public transportation as well as the general trend toward exercise&fitness provide a strong impetus for the choice of bicycle as transportation mode. While bicycles may not dominate as the preferred option for relatively long commutes (Bergström and Magnusson 2003, van Wee et al. 2006), especially in bad weather (Nankervis 1999) or when safe bike routes are absent, there is a clear case in favor of bicycles for short commutes along safe and easy bike routes under good weather conditions. It therefore comes as no surprise that even McDonald's has been rumored to have developed a take-out bag especially for bicyclists (adsoftheworld.com/media/dm/mcdonalds_mcbike).

While inclement weather and other bad riding conditions (e.g., hills, potholes, lack of bike routes) may dissuade bicyclists from riding their bicycles, safety concern is a major deterrent. Cyclists have been shown to be more prone to accidents in mixed traffic (Pucher and Dijkstra 2000). Surprisingly, as the number of cyclists goes up, cyclist fatality rate goes down (Pucher and Buehler 2008). Findings from another study (Kaplan and Prato 2015) illustrate how encouraging cycling increases safety through *safety in numbers*, and increasing safety in turn convinces even more people to cycle. Given these results, there is a strong incentive to get more cyclists on the road.

© Springer International Publishing AG 2016
R. Doss et al. (Eds.): FNSS 2016, CCIS 670, pp. 172–186, 2016.
DOI: 10.1007/978-3-319-48021-3_12

With more bicyclists come more and disparate kinds of bicycle riding needs. Bicycle manufacturers cater to the different needs and wants of bicyclists with the provision of a range of types such as light-weight, recumbent, off-road, etc. with options that include battery-assistance, among others. Other than the materials used as well as their quality and characteristics, the general setup on the bicycle has remained the same for a long time (i.e., decades). This is about to change in the very near future when connected bicycles from several companies hit the market on a large scale.

There are several initiatives toward that end, mostly from new companies that attempt to satisfy market demand for bicycles that not only provide their basic functionality, but also incorporate components that facilitate improvements in overall safe and pleasant bicycling experience. These initiatives are a direct result of potential demand as well as advances in related sensor-based enabler technologies with associated functionalities that are incorporated in relatively small form factors for seamless fit with(in) a bicycle frame. These sensor-incorporated bicycles are designed to provide real-time feedback to the bicycle rider as well as other entities (e.g., automobiles, infrastructure) nearby with the ultimate goal of ensuring safety to everyone while also providing a certain level of convenience to the bicycle rider and those nearby.

Mechanism for such instantaneous feedback did not exist in earlier mass-marketed bicycle models. In addition to real-time status information on the bicycle and its components (e.g., tire pressure, wear level of brake pads), existing technologies (e.g., Internet connection, social media, GPS) allow for the provision of information on surrounding areas (e.g., shortest route to destination, social network friends nearby, vehicle approaching close from behind) as well as trigger necessary and appropriate alerts (e.g., slow tire pressure loss, message friends and/or relatives when involved in an accident). On-board Internet access allows the bicyclist as well as the connected bicycle to communicate with other entities (e.g., other connected bicycles, infrastructure, social media) and people.

Advances in IoT (Internet of Things) technology, especially those related to sensor-technology, facilitate the incorporation of more useful features in the bicycle. IoT with sensors are found in a wide variety of applications that include transportation control and connected automobiles. Connected automobiles are ahead of the curve (vs. connected bicycles) in terms of IoT technology utilization. Given that connected automobiles and connected bicycles occupy and share similar space in terms of transportation, it is possible to learn from the experience on implementation and use of IoT technologies in connected automobiles. Clearly, given the overall footprint and other resource constraints, not all IoT applications in automobiles directly map to those in bicycles. Nevertheless, opportunities for technology transfer as well as new applications that are specific to bicycles can and should be identified and implemented for an overall improved and safe experience for the bicyclists as well as others (e.g., pedestrians, those in automobiles nearby).

The purpose of this paper is to study IoT in connected bicycle ecosystems. Specifically, we consider the synergies associated with connected bicycles, connected self, IoT, and social media. The remainder of this paper is organized as follows: We provide a brief discussion on IoT in the next section. This is followed by relevant discussion on quantified self in Sect. 3. With these background knowledge, we then discuss connected bicycles and some of their features that render them the 'connected' status in Sects. 4 and 5. In addition to those features that are available or expected to be available in connected bicycles, we also discuss other possible features that may be readily implemented. We conclude with a brief discussion on connected bicycles and its future in Sect. 6.

2 IoT

Gartner (http://www.gartner.com/it-glossary/internet-of-things/) defines IoT as *"the network of physical objects that contain embedded technology to communicate and sense or interact with their internal states or the external environment."* Essentially then, IoT comprises generally, but not necessarily, tiny intelligent entities (things) that are always 'connected' in the sense that they have the capability to communicate with other entities at all times. In addition to the range of entities that can be connected this way, the communication content as well as the ability to be 'on' at all times is a distinguishing feature of IoT. Clearly, the *communication content* depends on the device and its application. For example, albeit cliché, a coffeepot (wirelessly) connected to an alarm clock could turn itself on and brew coffee when the alarm goes off. A thermostat connected to motion sensors can automatically adjust the room temperature based on the time of the day, day of the week, and the presence of occupants and their overall desired temperature settings. Similar automation is achieved in a myriad of applications such as the adjustment of light intensity, ambient music, window shade, etc. before the guest walks into a hotel room, automated adjustment of gym machines when the next user walks close to it, notification of tire pressure to the car's computer, among others.

The Internet of Things (IoT) in its various forms is here, and has witnessed steady growth over the past several years. As a single thread that directly or indirectly connects several aspects of our lives, the combination of products and services with intelligent capabilities that comprise IoT has the potential to radically change the information and communication technology (ICT) landscape. IoT applications span a wide application area and include such disparate entities as wearable fitness trackers, connected automobiles, biometric passports, public transport tickets, among others.

The IoT revolution follows that of Wi-Fi of more than a decade ago and of smartphones of less than a decade ago in facilitating the addition of even more entities on wireless networks. The result is that disparate entities communicate with one another to seamlessly accomplish tasks and solve problems in real-time. IoT is somewhat of a misnomer since not all IoT devices are directly connected to the Internet. These devices operate through simple wireless protocols, and may be indirectly connected to be controlled through the Internet. Moreover, while the commonly visible usage of IoT devices (e.g., IoT thermostat) involve sensors in some form, IoT devices do not all have associated sensors. For example, among all currently used IoT devices, RFID (Radio-Frequency IDentification) tags comprise a dominant majority. Most RFID tags in use now are passive tags with applications in retail as well as pharmaceutical supply chains. Passive RFID tags do not have an on-board or associated sensor.

While a paired set of IoT devices may prove to be useful (e.g., coffee machine and alarm clock), the power of IoT comes with the seamless integration of several of these devices that work in consort to form a coherent system that is constantly on the look-out to identify and automatically address issues or pro-actively tend to the needs of its environment. Given that these IoT devices are generally a part of a network, there clearly exist synergies and associated network effects.

According to Gartner (2013), in 2009 the number of connected devices with unique IP addresses and those that were connected to the Internet were about 2.5 billion, mostly made up of cellphones and personal computers. They estimate that in 2020, the number of connected devices with unique IP addresses will increase to 30 billion. In addition to their industrial applications (e.g., manufacturing, supply chains, inventory management), IoT devices deliver a new level of convenience to our everyday lives. The primary drawback of such a setup is the increase in reliance on a potentially vulnerable ecosystem. As more objects in our everyday lives such as refrigerators, thermostats, and baby monitors are connected to general online IoT ecosystem, each such device represents yet another point of online exploitation for adversaries.

The potential impact of such an ecosystem on personal privacy and security cannot be overstated. Recent exploits such as Stuxnet, Flame, and Regin have raised issues related to industrial espionage in the era of connected devices (Osborne 2015). As more in number as well as variety of IoT devices are introduced, attacks targeted at the diversity of such newly connected endpoints will invariably become more common. There are two sides to such developments: while the developers envision a wealth of new capabilities, adversaries are offered a vast new attack surface.

All IoT devices have the potential to generate a huge amount of data, which must be handled carefully to avoid privacy and security related issues. With the increase in the number of IoT devices that pervade every facet of our lives, unlawful surveillance or intrusion into private life must be given serious consideration.

3   The Quantified Self

Quantified self refers to an individual who measures and tracks data about oneself, with specific reference to behavioral, biological, and physical data. In recent times, this has come to be associated with the regular use of fitness trackers to measure and keep track of one's biophysical metrics such as blood pressure, blood sugar, heart rate, sleep patterns, weight, among others. Such fitness trackers need not necessarily be wearables, and can include wi-fi connected bathroom scales, fitness tracking apps such as MapMyRun, etc.

A large number of quantified self-trackers are goal-oriented, with the ultimate goal of using health tracking data to initiate interventions that address issues related to cognitive alertness, depression, productivity, sleep quality, etc. A 2013 Pew Internet study found that 7 out of 10 adults in the US track one or more health metrics (Einarsen 2013). This study found that tracking is followed by appropriate actions such as a change in overall approach to health maintenance. This includes overall greater awareness of their health in being able to ask new questions of doctors and on how an illness is treated. The widespread use of healthcare-oriented online social networks (HOSN) has helped facilitate this process through shared experiences with related health concerns and associated community support (Sadovykh 2011).

With the growing number of devices that are specifically targeted at the Quantified Self (QS) phenomenon, Gartner (McIntyre 2014) forecasts the market for such devices to reach $5 billion by 2016. Recent advances in unobtrusive sensor technology such as those that measure heart rate, respiration, posture, etc. have helped with the growth in QS popularity.

In addition to such measurement and tracking by the users themselves, some of these QS sensors are used in various applications to automate interventions when exceptions are identified. For example, when more variable steering wheel movement in automobiles that signify drowsy or fatigued drivers is detected, appropriate interventions such as verbal alerts, seat vibration, etc. can be instantiated to help with the situation. Such real-time interventions can help reduce unsafe driving incidents and related accidents.

Real-time accident assistance through crash response technology is already being used in automobiles with the integration of QS sensors and cloud-based electronic medical records (EMR) data. (e.g., GM's OnStar Automatic Crash Response system that works in consort with GPS data, Europe's Automatic Emergency Call - eCall). When an accident occurs, automobile-based alert systems can transfer collected on-board sensor data to help assess the impact of the accident and to automatically contact the most appropriate first responders and trauma center based on the level of trauma. Such data can also be automatically transmitted to hospitals, ambulances, and law enforcement systems. In consort with EMR and the identity of the occupants of the vehicle, the most appropriate response can be determined and necessary help can be immediately dispatched.

QS sensors in automobiles can also be used to detect abnormal heart rhythms. For example Toyota's ECG steering wheel monitors the driver's heart rate, and automatically slows down the vehicle when it detects any abnormal heart activity (Hutchings 2011). A related application in automobiles is the use of this technology for tracking the health of the driver as well as the passengers when they use the car.

The measurement accuracy in QS devices is a concern, especially when used for serious purposes as opposed to recreational purposes where measurements are used only as a benchmark to be compared over time. While sensors in wearables are not necessarily accurate due to their form-factor, size, and other resource (e.g., power supply, memory, processing power) constraints, their counterparts in automobiles and other non-wearable sensors are reasonably accurate.

4   Connected Bicycle

The connected bicycle concept allows for the bicycle and/or its rider to have continuous Internet connection, which is then used to send data from IoT devices on-board or on the rider as well as receive data from external sources such as infrastructure or a vehicle. Such data can be seamlessly integrated in applications such as real-time traffic alerts, fatigue detection, real-time assistance for accidents, key-less authentication, digital identity verification, diagnostics, among others. For example, knowledge of a bicycle's location, speed, direction, and destination can be used to help avoid streets with congested automobile traffic. Connected bicycles with on-board GPS can provide traffic warnings, alert the riders to road hazards, improve overall traffic conditions, and communicate with other entities (e.g., automobiles) for collision avoidance.

While connected bicycle is a relatively new concept, connected automobiles have evolved over the years. The essential thrust in connected automobile systems is the automation of physical and cognitive tasks. It is important to note that unlike industries such as education, entertainment, financial services, health care, publishing, and music, the connected automobile industry has not been forced to reinvent itself and reorganize its entire ecosystem. However, unlike the simultaneous existence of digital and physical nature of products in these other industries, the automobile industry integrated advances in technology with minimal impact to its ecosystem.

A connected car can have several sensors to measure, record, and take appropriate action based on real-time information. Such sensors include biometric sensors that measure heart rate, etc. of the automobile's occupants as well as those that measure and maintain historical data on braking activity, speed, air pressure in tires, engine oil level, coolant level, any identified exceptions, etc. As a part of the largest connected world, the connected automobile is seamlessly connected with other entities on the Internet, presumably cognizant of privacy and security concerns related to the automobile as well as its occupants.

While automobiles are complex vehicles compared to bicycles, it is nevertheless possible to learn from the vast experience of connected automobile systems. The similarities between connected automobiles and connected bicycles include their use for transportation/recreation, the presence of on-board biometric and vehicle sensors, among others. Clearly, there are differences between connected automobiles and connected bicycles. Whereas there are several current initiatives toward the development and deployment of autonomous automobiles, we are not aware of similar initiatives for autonomous bicycles.

There are several advantages to connected bicycles. Detailed knowledge of bicycle data will necessarily lead to improved diagnostics, timely repair, and appropriate maintenance, resulting in a fundamental shift from reactive service to preventive and proactive maintenance. Such detailed data in consort with data analytics will enable predictive action that facilitates the identification and avoidance of issues before they happen or become serious. Reactive response after the fact is not useful in a majority of circumstances. Predictive action as well as real-time control helps alleviate the situation before damage is done.

5   Mass-Marketed Connected Bicycles

The number of connected devices/things in existence has already exceeded 16 billion, and is expected to reach 40.9 billion by 2020 according to ABI Research (ABI, 2014). With the addition of each such device, with or without associated sensors, is a concomitant increase in data flow. Cisco IBSG (2015) estimates the machine-to-machine (M2M) IP traffic to grow 15-fold from 308 petabytes in 2014 to 4.6 exabytes by 2019 which translates respectively to about 0.5 % to 2.7 % of global IP traffic. While the contribution of connected bicycles to these statistics is most likely minimal, the opportunities and associated impact on the general population cannot be ignored.

Over the past few years, at least a dozen companies have expressed interest or have (concept or proof-of-concept) prototypes of connected bicycles or components that enable any bicycle to become a connected bicycle with various features and characteristics. While less than a handful of these are already in the market, the remaining initiatives are expected to come to fruition within the next few years.

After an extensive search of available as well as soon-to-become-available connected bicycles, we now discuss the features that already exist in the market or as prototypes in the connected bicycle ecosystem. The list of connected bicycle companies or models mentioned in this paper is not meant to be exhaustive, although we are fairly confident that the features mentioned here are exhaustive from currently available connected bicycle perspective. For illustration, we considered connected bicycles from Bikespike (www.jebiga.com/bikespike-team-bikespike/), Canyon (www.canyon.com), COBI (cobi.bike), Connected pedal (connectedcycle.com), Dubike (dubike.baidu.com), Helios bar (www.ridehelios.com), LOCK8 (lock8.me), MoDe (https://corporate.ford.com/microsites/sustainability-report-2014-15/doc/sr14-modeme-ebike-eu.pdf), Stromer (www.stromerbike.com), Vanhawks (www.vanhawks.com), Visiobike (www.visiobike.com), and Wi-MM (https://angel.co/wi-mm). We do not differentiate between built-in or brought-in connections - those that are marketed as connected bicycles and components (e.g., handlebar, pedal) that enable any bicycle to become connected.

5.1   GPS

There are very few, if any at all, connected bicycles that lack on-board GPS. These GPS devices are used primarily for identification of current location. GPS data from these devices are also recorded in some connected bicycles to help track/trace the bicycle's location. Such data are also used for navigation control, fitness-related applications, as well as to train and track performance of the bicyclist. Although the application and use of generated GPS data are somewhat different across connected bicycles, all the connected bicycles we considered have GPS. For example, the concept connected bicycle from Canyon uses data from the integrated GPS unit to track training sessions, trace the bicycle if/when it gets stolen, trigger emergency services call along with the location of the incident in case of an accident. The Helios bar uses GPS data to generate information on local weather, congestion in surrounding or intended path, parking, available public transportation nearby, and traffic conditions.

Several of the connected bicycles we considered have the feature that triggers an alert in case of an accident. The connected bicycles in Europe use the eCall (Emergency Call) service that is meant for rapid assistance to motorists involved in a collision. In the US, General Motor's OnStar Automatic Crash Response system works in consort with GPS data.

From a connected bicycle perspective, several concerns arise with the use of GPS for location identification. Battery life is an issue with GPS use. Accuracy level is also an issue (e.g., Lindsey et al. 2013). However, a significant issue is the ease with which the usefulness of GPS unit is disabled. The ready availability of GPS-jammers that interfere with GPS radio signal, GPS spoofing devices that send fake radio signal associated with a fake location, and mobile phone jammers that jam wireless signals when wireless carriers are used to send GPS information renders it possible for adversaries to effectively 'disable' the GPS functionality (to, for example, steal the bicycle). Metal shield (e.g., brass mesh) can also be used to shield GPS signals. Given these, the claim that the last location before GPS-disable is known, and therefore somehow the bicycle can be retrieved, may be a difficult sell.

5.2   Connectivity

Connectivity is what imparts a bicycle its 'connected' quality. This distinguishing feature is therefore of paramount importance for connected bicycles. It should go without explicitly stating that all connected bicycles we considered have some form of connectivity that include either wi-fi or the use of some form of cellular connection for communication with other entities (e.g., other connected bicycles). For example, the Canyon connected bicycle uses GPS in conjunction with GSM to track the bicycle when it's stolen. Although 2G GSM is scheduled to be phased out (e.g., Macau, Singapore), an appropriate means to connect the connected bicycle exists in connected bicycles. The Connected Pedal uses cellular connection through an integrated SIM (Subscriber Identity Module) card.

5.3   Anti-theft

Not all connected bicycles use the same anti-theft measure. The basic idea is to alert the owner or whoever is in charge of the bicycle when someone attempts to steal the bicycle so that an appropriate and timely response can be quickly instantiated. Such an alert can be triggered when the bicycle is physically moved from its parked location or when its position is moved which might signal a theft attempt. The COBI system uses an accelerometer to identify theft-in-progress and triggers a light and sound alarm to deter the theft from completion. The Vanhawks connected bicycle uses GPS-generated data along with a wi-fi direct module to trigger an alert.

Bikespike is a small GPS tracker device that is placed under a water-bottle cage along the bicycle frame. It begins to track when someone steals the bicycle, and provides regular (at user-specified frequency) GPS-generated updates on the current location of the bicycle. In a sense, this is not anti-theft but rather the provision of information that the bicycle owner can use to possibly locate the stolen bicycle. Similar to Bikespike, Wi-MM however is integrated into the water-bottle holder attached to the bicycle frame.

The digital smart lock Loc8 is designed to secure the bicycle using an app. With the use of Bluetooth, it detects when the bicycle rider walks away from the bicycle, and automatically activates the alarm. Loc8 uses an accelerometer, temperature sensor and smart cable to detect a theft attempt and triggers a 120 db alarm and simultaneously sends a text message to the bicycle owner/user. A distinguishing feature of Loc8 is the possibility for the community of Loc8-attached bicycles to police other bicycles that have Loc8 smart locks - when a stolen bicycle with Loc8 smart lock is nearby (another bicycle with Loc8 smart lock), this other bicycle relays the information (about location of the stolen bicycle) to appropriate parties to enable swift action and facilitate return of the bicycle to its rightful owner. Since the locking mechanism works through an app, the bicycle owner can also remotely unlock the bicycle to enable sharing (e.g., Piramuthu and Zhou 2016) of the bicycle with others. Loc8 also has GPS to track the bicycle when it's stolen in spite of the anti-theft mechanism in place.

The GPS-based theft alert system seems to be the preferred option in the connected bicycles we considered, although the use of GPS for this purpose has serious limitations especially from the perspective of a determined thief.

5.4   Blind-Spot Detection

Only one of the considered connected bicycles has this feature. The Valour model from Vanhawks uses ultrasonic means to detect the presence of other vehicles or objects in the blind spot and informs the rider.

Although not a connected bicycle, in 2013 Volvo introduced the 'Cyclist Detection' technology with the incorporation of a radar in the car's front grille and a rear-view mirror-mounted camera. When it detects a bicyclist or pedestrian on its path and the driver doesn't respond to its warning, it automatically applies the brake. Moreover, along with the Swedish engineering and design firm POC and the communications company Ericsson, the bicycle app Strava is used to share the position of car and bicycle to Volvo's cloud-based network. This information is used to determine if an accident seems imminent and to trigger an alert through the car's head-up display (HUD) as well as display a flashing signal in the bicyclist's helmet to warn both the car driver as well as the bicyclist. The ultimate goal is to eliminate blind spots between car and bicycle and to avoid related collisions between cars and bicycles.

Although not a blind-spot detector, the Visiobike has a rear-facing camera under the seat that records the last 3 min before an accident. This 'automatic accident detection' triggers emergency alerts to pre-specified recipients when the rider doesn't respond to 'Are you OK?' query displayed in the smartphone.

Ford's MoDe apparently has the ability to provide 'approaching car warning'. However, from publicly available information at the time of this writing, it is not clear as to how this functionality is accomplished.

5.5 Direction to Destination

Several connected bicycles use GPS technology to provide turn-by-turn directions to reach a chosen destination. In several of the cases, the bicycle rider is guided through the process with the use of haptic means where the technology is incorporated in the handlebar grips. Some of the connected bicycles have blinking lights that display the turn direction just before an intersection is reached. For example, the Dubike has a set of blinking arrow lights in the middle of the handlebar to show the direction. The COBI system uses GPS-generated information for navigation control. With help from GPS and Bluetooth technology, the Helios bar uses blinking lights to show the turn direction. All three MoDe e-bikes from Ford integrate with the rider's smartphone via the MoDe:Link app to deliver real-time turn-by-turn navigation.

5.6 Heart Rate Monitor

While heart rate monitors are ubiquitous in wearable fitness devices, it is not yet incorporated in a majority of connected bicycle designs. However, there are a few exceptions. For example, Ford's MoDe e-bikes incorporate a heart rate sensor. Data from this heart rate sensor is used in the 'no sweat' mode to dynamically modify the on-board motor's output to reduce physical effort and thereby the probability of perspiration of that bicycle's rider.

5.7 Bicycle Maintenance Status

Among the connected bicycles we considered, Canyon was the only one with the ability to provide real-time maintenance information on whether any moving or worn-out part such as brake pad, chain, or gear cable need replacement. With appropriate sensors that can be set to measure wear and tear of moving parts or parts that are prone to wear-out, it is quite conceivable to expect such sensors as well as related ones (e.g., tire wear, tire air pressure, brake disc wear, battery status) to be incorporated and become an integral part of connected bicycles. The connected bicycle can also include the feature whereby it sends a text message to its owner if a tire goes flat overnight. This can be readily accomplished through services such as ifttt (if this then that ifttt.com).

Knowledge of such maintenance information provides the bicycle rider with a quick real-time overview/snapshot of the bicycle's status at any point in time. In some instances, the rider gets enough lead time to take *predictive maintenance* action before it's too late. For example, when the chain wears or (brake or gear) cable wears down enough to affect effective performance, the bicycle rider has enough of a lead time to take appropriate action (i.e., replace the part) before damage occurs to other related components.

5.8   Pothole Detection

Ford's MoDe ebikes have this feature - of being able to detect potholes. However, there is no publicly available information (at the time of this writing) as to how this is accomplished. DuBike apparently also has this feature. However, it is accomplished in DuBike through social media when a bicyclist runs over a pothole and shares that information with all DuBike riders - an example of collaboration with a shared purpose. Vanhawks' Valour bicycles connects with other Vanhawks Valour bikes, and help improve frequently traveled routes over time to help the bicycle rider avoid congestion, potholes, and other pitfalls.

5.9   Social Media

With real-time connection to the social media, a related social media application or the connected bicycle can make suggestions on the next action (e.g., return bicycle to nearest service location, visit with a social media friend who is at a nearby service location, etc.). Connected bicycles can also use the social media to obtain timely help. For example, when a connected bicycle has a flat tire and the rider forgot to bring a spare tube or puncture patching took-kit, a nearby connected bicycle with the ability to help (i.e., has a spare bicycle tube of the same size or has a patch tool kit) can be informed of this through social media. When bicyclists who are willing and able to help exist, the rider or the bicycle that could use this help can be instantaneously informed of the availability of needed help. The two (help provider and help receiver) can then coordinate the process by arranging to meet at some convenient location nearby. In a similar vein, it is also possible to get advice from other bicyclists on issues related to one's connected bicycle in real-time through social media. To our knowledge, none of the connected bicycles considered include this feature yet.

6   Discussion, Future Trend, Conclusion

We considered the incorporation of IoT devices that provide *actionable intelligence* in mass-marketed connected bicycles. Some of these devices or functionalities (e.g., heart rate monitor) generate information related to the connected self. The seamless integration of such IoT devices and sensors work in consort to potentially improve the safety of the bicyclist as well as those in close physical proximity to this connected bicycle. This potential is multi-faceted and arises as a direct consequence of IoT-generated information for proper bicycle maintenance, the provision of early warning, and ensuring that the bicyclists heed warnings. What connected bicycles cannot yet do is control the safety *perception* of bicyclists in mixed-traffic conditions, especially when there are relatively few bicyclists on the road. Extant research points toward higher safety perception in the presence of a large number of bicyclists. Perhaps the ratio of number of bicycles and automobiles on a mixed-traffic road has a propensity to follow the *Power Law* (Hatton et al. 2015), and can be nudged higher by encouraging more bicyclists.

As the number of players in this domain increases, there is a convergence in the set of common core functionalities that are being offered in connected bicycles from different vendors. It is also of interest to note that almost all connected bicycle firms are relatively new start-ups, with the curious absence of established bicycle firms.

Connected automobiles include a large number of on-board computers that control almost all main functionalities. Whereas connected bicycles are a lot less complex (vs. connected automobiles) in several dimensions, it is possible for technology transfer (e.g., blind-spot detection) between the two connected transportation modes despite physical/resource constraints in bicycles. Clearly, the set of functionalities that are available in connected bicycles is only bound to increase in number with advances in related technology, synergy with existing functionalities, as well as the demand for such functionalities. We conclude with brief discussions on a few functionalities that are not yet available in connected bicycles, but could be readily incorporated.

6.1   Demand-Actuated Traffic Signal

Traffic signals operate through inductive loop sensors, video, magnetometer, or digital radar. Oftentimes, given the low footprint and profile of bicycles, these signal systems fail to register the bicycle's presence. Perhaps connected bicycles can include some IoT means to aid traffic signals sense their presence.

6.2   Traffic-Related Air Pollution Detection

Some of the deleterious effects of traffic-related air pollution on bicyclists' health are known (e.g., Peters et al. 2014, Weichenthal et al. 2011). Busy automobile traffic streets are generally observed to have very poor quality air with high concentrations of soot and ultra-fine particles. While it is preferable for bicyclists to use bicycle routes that are farther away from such busy traffic, the bicyclist is oftentimes forced to take the busy traffic route due to lack of better options. Given this state of affairs, bicycle-mounted real-time personal pollution monitors (e.g., Bales et al. 2014, Kanjo and Lanshoff 2007) empower bicyclists when there is a choice on bicycle routes to take based on time of day and day of week.

6.3   Continuous Blood Glucose Monitoring (CGM)

For diabetic bicycle riders who require continuous glucose monitoring (CGM), a traditional option is through invasive means where a finger-prick test is done as often as is necessary or a sensor is placed under the skin. Accurate glucose level knowledge is of utmost importance when dosing insulin for some diabetics, and the development of a noninvasive and wearable monitor would allow for continuous glucose level readings whereby the reader is instantly alerted when intervention is needed.

Such a device is under development by several firms (e.g., glucosense - www.glucosense.net, dexcom - www.theverge.com/2015/8/12/9136743/ google-dexcom-cloud-connected-disposable-glucose-monitor). In the future, as more progress is made in this area, it is possible for such a device to be seamlessly incorporated in a connected bicycle.

6.4   Fatigue Detection

Fatigue is implicated in 15–20% of road (http://www.rospa.com/road-safety/ advice/drivers/fatigue/road-accidents/) and air (https://www.eurocockpit.be/ pages/fatigue-in-accidents) accidents. Early warning signs of fatigue such as slower heart rate and breathing rate can be detected by sensors in connected bicycles, which can alert (e.g., haptic) the bicycle rider.

6.5   Geosense

Geosense has been gaining popularity in the connected automobile ecosystem. With the availability of GPS systems in connected bicycles, it is feasible to incorporate geosense-based applications in these bicycles. Such an application can then provide unsafe neighborhood alert, speed limit alert (especially useful when the speed limit is in the lower ranges and below that which is easily reached by bicycles), among others. Recorded geosense information can also be used to show as evidence (of bicycle as the means to commute to work) for automobile insurance purposes.

Although its usefulness is sometimes questionable, context-aware services (e.g., traffic warning message delivery, user-proximity-based advertising) can also be readily implemented in connected bicycles.

References

ABI Research. The Internet of Things will Drive Wireless Connected Devices to 40.9 Billion in 2020 (2014). https://www.abiresearch.com/press/ the-internet-of-things-will-drive-wireless-connect/

Bales, E., Nikzad, N., Ziftci, C., Quick, N., Griswold, W., Patrick, K.: Personal pollution monitoring: mobile real-time airquality in daily life. Technical report, UCSD Computer Science Department (2014)

Bergström, A., Magnusson, R.: Potential of transferring car trips to bicycle during winter. Transp. Res. Part A **37**(8), 649–666 (2003)

Bridgelall, R.: Inertial sensor sample rate selection for ride quality measures. J. Infrastruct. Syst. **21**(2), 04014039:1–04014039:5 (2015)

Capital Bikeshare: Capital Bikeshare (2014). http://cabidashboard.ddot.dc.gov/

Cisco IBSG: The zettabyte era: observation of strains: trends and analysis. White Paper, CISCO Virtual Networking Index, May 2011. http://www.cisco.com/c/en/us/solutions/collateral/service-provider/visual-networking-index-vni/VNI_Hyperconnectivity_WP.html

Eckhoff, D., Sommer, C.: Driving for big data? Privacy concerns in vehicular networking. IEEE Secur. Privacy **12**(1), 77–79 (2014)

Einarsen, B.: 7 in 10 US adults track a health indicator. In: Klick Health (2013). http://www.klick.com/health/news/blog/insights/7-in-10-us-adults-track-a-health-indicator/

Gartner: Gartner says it's the beginning of a new era: the digital industrial economy. In: Gartner Symposium/ITxpo, Orlando, 6–10 October (2013). http://www.gartner.com/newsroom/id/2602817

Hatton, I.A., McCann, K.S., Fryxell, J.M., Davies, T.J., Smerlak, M., Sinclair, A.R.E., Loreau, M.: The predator-prey power law: biomass scaling across terrestrial and aquatic biomes. Science **349**(6252), aac6284 (2015)

Hutchings, E.: Toyota's ECG steering wheel monitors your heart rate as you drive. In: PSFK Innovation (2011). http://www.psfk.com/2011/08/toyotas-ecg-steering-wheel-monitors-your-heart-rate-as-you-drive.html#!bOj0Qx

Kanjo, E., Lanshoff, P.: Mobile phones to monitor pollution. IEEE Distrib. Syst. Online **8**(7), art no. 0707–07002 (2007)

Kaplan, S., Prato, C.G.: A spatial analysis of land use and network effects on frequency and severity of cyclist-motorist crashes in the Copenhagen region. Traffic Inj. Prev. **16**(7), 724–731 (2015)

Kohler, W.J., Colbert-Taylor, A.: Current law and potential legal issues pertaining to automated, autonomous and connected vehicles. Santa Clara High Tech. Law J. **31**(1), 99–138 (2015)

Lindsey, G., Hankey, S., Wang, X., Chen, J., Gorjestani, A.: Feasibility of using GPS to track bicycle lane positioning. Technical report #: CTS 13–16. University of Minnesota, ITS Institute (2013)

McIntyre, A.: Market Trends: Enter the Wearable Electronics Market With Products for the Quantified Self (2014). https://www.gartner.com/doc/2537715/market-trends-enter-wearable-electronics

Nankervis, M.: The effect of weather and climate on bicycle commuting. Transp. Res. Part A **33**(6), 417–431 (1999)

Osborne, C.: Beyond Stuxnet and Flame: Equation 'most advanced' cybercriminal gang recorded, ZDNet, 16 February 2015. http://www.zdnet.com/article/beyond-stuxnet-and-flame-equation-group-most-advanced-cybercriminal-gang-recorded/

Peters, J., Bossche, J.V.D., Reggente, M., Poppel, M.V., Baets, B.D., Theunis, J.: Cyclist exposure to UFP and BC on urban routes in Antwerp, Belgium. Atmos. Environ. **92**, 31–43 (2014)

Piramuthu, O.B., Zhou, W.: Bicycle sharing, social media, and environmental sustainability. In: Proceedings of the 49th Hawaii International Conference on System Sciences (HICSS-49), pp. 2078–2083. IEEE Computer Society (2016)

Pucher, J., Dijkstra, L.: Making walking and cycling safer: lessons from Europe. Transp. Q. **54**(3), 25–50 (2000)

Pucher, J., Buehler, R.: Making cycling irresistible: lessons from the Netherlands, Denmark, and Germany. Transp. Rev. **28**(4), 495–528 (2008)

Sadovykh, V.: Decision making in online social networks. MCom thesis, Information Systems, University of Auckland (2011)

Shaheen, S.: Introduction to shared-use mobility: definitions, trends, and understand. In: Shared-Use Mobility Summit. Transportation Sustainability Research Center, UC Berkeley, Washington, D.C. (2014)

van Wee, B., Rietveld, P., Meurs, H.: Is average daily travel time expenditure constant? In search of explanations for an increase in average travel time. J. Transp. Geograph. **14**(2), 109–122 (2006)

Weichenthal, S., Kulka, R., Dubeau, A., Martin, C., Wang, D., Dales, R.: Traffic-related air pollution and acute changes in heart rate variability and respiratory function in urban cyclists. Environ. Health Perspect. **119**(10), 1373–1378 (2011)

Collaborative Network Coding in Opportunistic Mobile Social Network

Tzu-Chieh Tsai, Chien-Chun Han[(✉)], and Shou-Yu Yen

Computer Science Department, National Chengchi University, Taipei, Taiwan
ttsai@cs.nccu.edu.tw, hanjord@gmail.com, edwardyen307@gmail.com

Abstract. Opportunistic mobile social network is a type of delayed tolerant network, where nodes with mobility contacts each other through short range wireless communications. Recently, many related applications are emerging, such as Firechat. However, message dissemination in opportunistic mobile social network is a challenging task. We propose collaborative network coding that enables users to take part in improving the performance of using network coding for message dissemination. The proposed method is evaluated by trace data conducted by participants who may not know each other in advance for a more realistic simulation of real world opportunistic mobile social network. Simulation result shows that our proposed method out performs flooding based message dissemination.

Keywords: Opportunistic mobile social network · Network coding · Mobile social network · Delay tolerant network

1 Introduction

Mobile social network (MSN) is a special network formed by mobile nodes communicating with short distance wireless communication capability, such as Bluetooth or WiFi [2].

Recently, there are emerging applications based on mobile social network. Firechat [3] enables instant messaging without requiring Internet access, which is a powerful solution in some special scenarios when Internet is unavailable. Happn [4] is a mobile social service that let users get to make new friends that he or she often physically "encounters".

Such mobile social network is connected in an opportunistic manner. However, communication in opportunistic mobile social network is a very difficult task. Inter connectivity formed by short distance wireless communication can open up powerful new possibilities, such as communication without infrastructure, proximity, and recording physical encounters between nodes, but the network suffers from constant disconnection due to nodes' mobility. Other challenges also includes limited resources on mobile devices, such as limited energy and computing power.

Network coding is a promising solution for mobile social network. Benefits of network coding include improved robustness of network operations, higher energy efficiency in wireless radios, and better security [5]. Network coding requires

© Springer International Publishing AG 2016
R. Doss et al. (Eds.): FNSS 2016, CCIS 670, pp. 187–194, 2016.
DOI: 10.1007/978-3-319-48021-3_13

wireless broadcasting. On mobile devices, using Wifi Direct and Bluetooth Low energy can leverage the advantage of network broadcasting without setting up an infrastructure.

We propose a protocol based on network coding to enable multi-hop instant messaging in mobile social network. Our protocol is specially designed to leverage the advantage of Bluetooth Low Energy, which is a widely adapted technology in mobile devices on the market.

Our instant messaging service lets users create chatrooms, and automatically invites new users to the chatroom using recommendation based on potential social similarity.

2   Related Work

Opportunistic mobile social network is a kind of delay tolerant network with social relations between the nodes. Network coding have been proposed for delay tolerant network in [9], shown to improve network performance compared to without using network coding. [5] compared network coding with simulation and running applications on real devices, and founded the results match, indicating that simulation can study larger scale of delay tolerant network with network coding at a much more manageable cost. [14] considers message delivery with hard deadlines while using network coding when broadcasting messages by constructing broadcasting trees, and showed that transmission number is reduced with a good delivery ratio.

3   Collaborative Network Coding

We target instant messaging in opportunistic mobile social network. Inter communication between nodes are formed by Bluetooth Low Energy, to save energy consumption while providing adequate performance.

1. **Multi Hop Enabled Instant Messaging**

   A user can create a chatroom for others to join. After a user creates his or her chatroom, as he or she moves around, the user will physically contact with other users. We calculate the social similarity of users using social similarity [6]:

   $$S_{i,j}(t) = 1 + |F_i(t) \cap F_j(t)|$$

   Where $F_i(t), F_j(t)$ is the set of friends of node i, j at time t. If node i has higher $S_{i,j}(t)$ for j, it shares more common friends with j, and is more likely to be interested in meeting with each other. Note that the friends in the above equation is calculated from most recent contacted users.

   When node i contacts with node j and node i holds a chatroom or is following a chatroom, node i can calculate $S_{i,j}(t)$ to determine whether to invite j to the chatroom.

2. Collaborative Network Coding

We propose a message dissemination protocol based on network coding and collaboration between users. Network coding have been studied in the field of delay tolerant network, and have been proved effective [7]. In our protocol, the network coding part is based on random linear network coding. Messages are transferred in a store and forward manner, while in each transmission, messages are broadcasted to a node's neighbors.

a. Random Linear Network Coding

In random linear coding, packets are composed of encoding vector and payload vector. Encoding vector shows how packets in the payload vector can be combined. Messages are combined by random linear combinations of encoding vectors and payload vectors. When a node receives messages, it tries to recover the original data by performing Gaussian elimination.

The dimension of the information vector is the length of the payload M, while the dimension of the encoding vector is the maximum number of original messages that can be combined, denoted as m. However, large m will generate high computation overhead. To deal with the high dimension issue, packets can be divided into generations [8]. [9] Shows that the size of the generation is crucial to the performance of using network coding in delay tolerant network.

Since our scenario is instant messaging, we divide generations by using a period of time, i.e. 1 h. Messages generated in the period will belong to the corresponding generation.

b. Collaborative Random Linear Network Coding Protocol

During a contact between nodes, contact time is limited due to the mobility, a node can only transfer a part of the messages in its message queue. Therefore, we would like to enhance the performance of random linear network coding by trying to transfer messages that will be more likely to benefit the decoding of messages for the recipient. When a node broadcasts messages to the recipients, the recipients will ACK the current messages they have already decoded to the broadcaster, and the broadcaster will select messages to be broadcasted in the next round based by the ACK. To try to maximize the benefit of broadcasting in next round, the broadcaster will broadcast messages that are most likely to be needed by the recipients.

Each node manages three data structures that stores messages. First and second of them are messages generated by the node M_{Self}, and messages decoded by the node $M_{Decoded}$, which are all decoded messages that can be accessed by any application running on the node's mobile device. The third data structure holds the received coded messages. Therefore, M_{Coded}. $M_{Self}, M_{Decoded}, and M_{Coded}$ are candidate messages to be broadcasted to a node's neighbors, denoted as $C_1, C_2 \ldots C_n$, where n = total messages cached in a node's buffer.

When a node i encounters other nodes $S_1, S_2 \ldots S_k$, it will initially randomly select some messages from and preforms broadcast of the messages to its neighbors.

The neighbors will send Ack back to node i, containing the identifiers of messages they have already decoded.

To broadcast messages that are more likely to benefit node i's neighbors in decoding their coded messages, we form a benefit matrix (Fig. 1):

$$
\begin{array}{c}
\begin{array}{ccc} S_2 & S_3 & \quad\quad S_k \end{array} \\
\begin{array}{c} C_1 \\ C_2 \\ \\ C_h \end{array}
\begin{bmatrix}
a_{00} & a_{01} & & a_{0k} \\
a_{10} & a_{11} & \cdots & a_{1k} \\
& \vdots & \ddots & \vdots \\
a_{h0} & a_{h1} & \cdots & a_{hk}
\end{bmatrix}
\end{array}
$$

Fig. 1. Benefit matrix

Where $a_{hk} = 1$ of S_k do not have C_h.

We then can calculate how C_h may benefit nearby neighbors, denoted as Gain (C_h):

$$\mathrm{Gain}\,(C_h) = \sum a_{hk}$$

Messages in C_h are ordered by Gain (C_h), and higher Gain (C_h) will result in a higher broadcasting priority. The protocol is described in the Fig. 2 below:

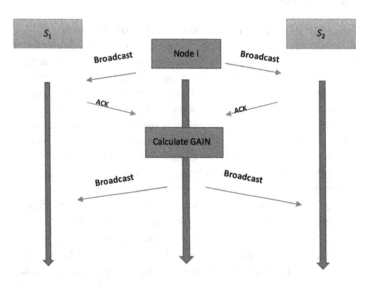

Fig. 2. Protocol flow

4 Simulation Settings

4.1 Simulation Environment

We use The ONE simulator [10] to conduct simulation and evaluation for this research. The ONE simulator does not support network coding by default, we extended The ONE to support network coding referencing [15].

Our target scenario is to enable multi hop instant messaging in mobile social network, and let users meet new friends that they are likely to be interested in meeting with. Therefore, the trace data we use must include users that may or may not know each other in advance. Existing trace data such as MIT reality [12], Cambridge [13] and InfoComm [14] are all collected from participants of same group, i.e. same department of students or same users participating in a same event. We choose the trace data in our previous work [11], NCCU Trace, as the mobility data.

The NCCU Trace is collected by mobility traces of students from several departments in National Chengchi University. The participants do not know each other in advance. There were 115 available user data in our trace data in total, and the experiment lasted for two weeks, from 17th Dec to 31st Dec in 2014.

4.2 Simulation Setting

The simulation settings are listed in the Table 1 below:

Table 1. Simulation settings

Area	6750*5100 m
Simulation time	172800 s
Transmission range	20 m
Buffer size	250 messages
Number of chatrooms	10
Total message created	12559

Since our target is Bluetooth Low Energy, we set the radio to 20 m. The simulation time is two days per round. Messages are created by nodes in its corresponding chatrooms, and only nodes who follows the chatrooms will receive the messages. Node only receive messages generated after it joins the chatroom.

4.3 Simulation Results

4.3.1 Delivery Rate

Messages delivery rate is calculated by counting the percentage of messages delivered to the nodes following the chatrooms (Fig. 3).

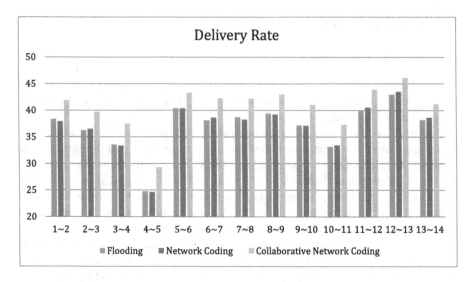

Fig. 3. Delivery rate

Simulation shows that Collaborative Network Coding out performs flooding based message broadcasting by about 10 %, while using only network coding with randomly picking candidate messages to deliver performs closely to flooding.

4.3.2 Message Delivery Delay

Message delay is calculated by comparing the message created time and message delivered time to all nodes following the corresponding chatroom (Fig. 4).

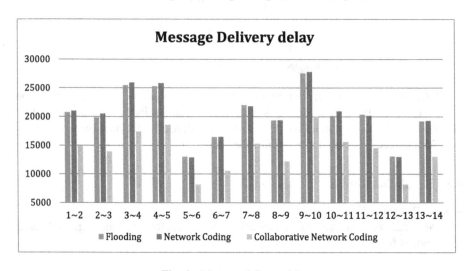

Fig. 4. Message delivery delay

It is observed that Collaborative Network Coding greatly out performs flooding in terms of message delivery delay, indicating that out proposed method is a more ideal solution for message chat in opportunistic mobile social networks.

4.3.3   Overhead
The transmission overhead is defined as (Fig. 5):

$$\frac{Transmitted messages - Received messages}{Received messages}$$

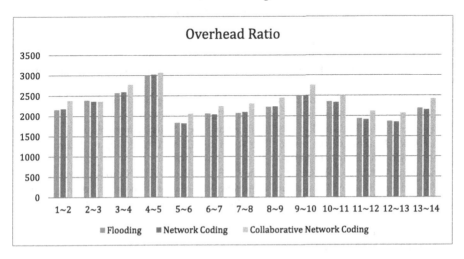

Fig. 5. Overhead ratio

From the result above, flooding, network coding, and Collaborative Network Coding performs closely in terms of transmission overhead. Since Collaborative Network Coding out performs flooding and network coding in terms of delivery ratio and delay, transmission overhead cannot be improved significantly.

5   Conclusion and Future Work

We have proposed a enhanced network coding protocol by leveraging collaboration of users in opportunistic mobile social network. Users are automatically invited to chatrooms according to their social similarity, people who have more common friends will join the same chatroom.

Simulation result shows that our proposed protocol is effective and out performs flooding based message broadcasting in terms of message delivery ratio and message delivery delay, while maintaining acceptable overhead.

References

1. Chin, A., Zhang, D. (eds.): Mobile Social Networking: An Innovative Approach. Computational Social Sciences, vol. XIV, 243 p., 64 illus. 61 illus. in color (2014). ISBN: 978-1 4614-8579-0. doi:10.1007/978-1-4614-8579-7
2. Lu, Z., Wen, Y., Cao, G.: Information diffusion in mobile social networks: the speed perspective. In: IEEE INFOCOM (2014)
3. http://opengarden.com/firechat/
4. https://www.happn.com/
5. Chen, Y., et al.: Delay-tolerant networks and network coding: comparative studies on simulated and real-device experiments. Comput. Netw. **83**, 349–362 (2015)
6. Zhu, K., Li, W., Fu, X.: Rethinking routing information in mobile social networks: location-based or social-based? Comput. Commun. **42**, 24–37 (2014)
7. Lin, Y., Li, B., Liang, B.: Efficient network coded data transmissions in disruption tolerant networks. In: INFOCOM 2008, IEEE the 27th Conference on Computer Communications. IEEE (2008)
8. Chou, P.A., Wu, Y., Jain, K.: Practical network coding. In: Proceedings of Allerton Conference on Communication, Control, and Computing (2003)
9. Widmer, J., Le Boudec, J.-Y.: Network coding for efficient communication in extreme networks. In: Proceedings of the 2005 ACM SIGCOMM Workshop on Delay-Tolerant Networking, pp. 284–291 (2005)
10. Keränen, A., Ott, J., Kärkkäinen, T.: The ONE simulator for DTN protocol evaluation. In: Proceedings of the 2nd International Conference on Simulation Tools and Techniques, ICST (Institute for Computer Sciences, Social-Informatics and Telecommunications Engineering), p. 55 (2009)
11. Tsai, T.-C., Chan, H.-H.: NCCU trace: social-network-aware mobility trace. IEEE Commun. Mag. **53**(10), 144–149 (2015)
12. Eagle, N., Pentland, A.: Reality mining: sensing complex social systems. Pers. Ubiquitous Comput. **10**(4), 255–268 (2006)
13. Hui, P.: People are the network: experimental design and evaluation of social-based forwarding algorithms. Ph.D. dissertation, UCAM-CL-TR-713. University of Cambridge, Computer Laboratory (2008)
14. Srinivasan, V., Motani, M., Ooi, W.T.: Analysis and implications of student contact patterns derived from campus schedules. In: Proceedings of ACM MobiCom, Los Angeles, CA, pp. 86–97, September 2006
15. Ostovari, P., Khreishah, A., Wu, J.: Broadcasting with hard deadlines in wireless multihop networks using network coding. Wirel. Commun. Mob. Comput. **15**(5), 983–999 (2015)
16. http://lokeller.github.io/ncutils/

Author Index

Printed in the United States
By Bookmasters